The Cosmic Panorama: ProtoCosmos, HyperCosmos, Unified SuperStandard Theory (UST) Derivation

Stephen Blaha Ph. D.
Blaha Research

A Research Publication

Pingree-Hill Publishing

MMXXII

ISBN: 979-8-9858266-3-0

Rev. 00/00/01 October 29, 2022

To Margaret

Some Other Books by Stephen Blaha

All the Megaverse! Starships Exploring the Endless Universes of the Cosmos using the Baryonic Force (Blaha Research, Auburn, NH, 2014)

SuperCivilizations: Civilizations as Superorganisms (McMann-Fisher Publishing, Auburn, NH, 2010)

All the Universe! Faster Than Light Tachyon Quark Starships & Particle Accelerators with the LHC as a Prototype Starship Drive Scientific Edition (Pingree-Hill Publishing, Auburn, NH, 2011).

Unification of God Theory and Unified SuperStandard Model THIRD EDITION (Pingree Hill Publishing, Auburn, NH, 2018).

The Exact QED Calculation of the Fine Structure Constant Implies ALL 4D Universes have the Same Physics/Life Prospects (Pingree Hill Publishing, Auburn, NH, 2019).

Integration of General Relativity and Quantum Theory: Octonion Cosmology, GiFT, Creation/Annihilation Spaces CASe, Reduction of Spaces to a Few Fermions and Symmetries in Fundamental Frames (Pingree Hill Publishing, Auburn, NH, 2021).

Passing Through Nature to Eternity ProtoCosmos, HyperCosmos, Unified SuperStandard Theory (Pingree Hill Publishing, Auburn, NH, 2022).

HyperCosmos Fractionation and Fundamental Reference Frame Based Unification: Particle Inner Space Basis of Parton and Dual Resonance Models (Pingree Hill Publishing, Auburn, NH, 2022).

A New UniDimension ProtoCosmos and SuperString F-Theory Relation to the HyperCosmos (Pingree Hill Publishing, Auburn, NH, 2022).

Available on Amazon.com, bn.com Amazon.co.uk and other international web sites as well as at better bookstores (through Ingram Distributors).

CONTENTS

1. THE HYPERCOSMOS AND UNIFIED SUPERSTANDARD THEORY (UST) 1

1.1 SYNOPSIS OF THE HYPERCOSMOS AND THE UNIFIED SUPERSTANDARD THEORY 1
1.1.1 Unified SuperStandard Theory *2*
1.1.2 The Standard Model Theory *7*
1.1.3 Limos and Fractionation *7*
1.2 THE LOGIC OF THE DERIVATION OF THE UNIFIED SUPERSTANDARD THEORY 7
1.2.1 The ProtoCosmos Stage *7*
1.2.1.1 Generation of Spaces from an Energy 8
1.2.2 The HyperCosmos Stage *9*
1.2.3 The QUeST Stage *10*
1.2.4 The UST Stage *10*
1.2.5 The Standard Model Stage *10*

APPENDIX 1-A. EXPERIMENTAL SUPPORT FOR THE HYPERCOSMOS AND UST 11

2. UNIDIMENSION PROTOCOSMOS MODEL 13

2.1 NEW UNIDIMENSION PROTOCOSMOS MODEL 13
2.1.1 UniDimension Force *13*
2.1.2 ProtoCosmos Free Space Wave Function *14*

3. TOWER PROTOCOSMOS MODEL 19

3.1 ENERGY VALUES DETERMINED BY R 20
3.2 SPIN – ENERGY RELATION 20
4.2.1 Spin – Amplitude Relation *21*
3.3 UNIVERSE WAVE FUNCTION 22

4. THE HYPERCOSMOS TRAJECTORY SPACES AS REGGE TRAJECTORY RESONANCES 25

4.1 SPIN ANGULAR MOMENTUM DEPENDENCE OF THE PROTOCOSMOS TRAJECTORY 25
4.2 INTERMEDIATE SPACES OF THE PROTOCOSMOS TRAJECTORY 25
4.3 DECAY OF "RESONANCES" OF THE PROTOCOSMOS TRAJECTORY 26

5. UNIDIMENSION UNIVERSE WAVE FUNCTIONS 27

5.1 ROLE OF UNIVERSE WAVE FUNCTION 27
5.2 HARMONIC OSCILLATOR UNIVERSE BINDING POTENTIAL 27

6. DERIVATION OF THE HYPERCOSMOS SPACES FROM CREATION/ANNIHILATION OPERATORS 29

6.1 GIFT FOR PARTICLE CREATION/ANNIHILATION OPERATOR TRANSFORMATIONS 29
6.1.1 GiFT for Scalar Creation/Annihilation Operator Transformations *30*
6.1.1.1 The B_1 Transformation 32
6.1.1.2 The B_2 Transformation 32
6.1.2 States in the Respective Coordinate Systems *33*
6.2 SCALAR PARTICLE CREATION/ANNIHILATION SPACE (CASE) 33
6.3 GIFT FOR *FERMION* CREATION/ANNIHILATION OPERATOR TRANSFORMATIONS 34
6.3.1 Transformations of the b and b† Operators without Spin Mixing or b – d Mixing *35*
6.3.2 Transformations of the d and d† Operators without Spin Mixing or b – d Mixing *36*

6.4 OPERATORS OF TYPES B AND D *WITH* SPIN MIXING AND B – D MIXING 36
6.6.1 Four Space-Time Dimension Case 37
6.4.2 Higher Space-Time Dimension Case 38
6.4.3 Size of the Fundamental Representations of su($2^{r/2}$, $2^{r/2}$) 38
6.5 THE HYPERCOSMOS SPECTRUM OF SPACES *DERIVED* 39
6.6 THE FERMION CASE GROUPS 39
6.7 THE N > 9 SPECTRUM SPACES OF NEGATIVE SPACE-TIME DIMENSION 40
6.8 ANTI-SPACES AND ANTI-UNIVERSES 40

APPENDIX 6-A. TWO-TIER AND PSEUDOQUANTUM THEORY FORMALISMS 41

THE PSEUDOQUANTUM THEORY OF COLOR CONFINEMENT APPEARS IN 41
6-A.1 REASONS FOR TWO-TIER QUANTUM THEORY 42
6-A.2 TWO-TIER FEATURES IN 4-DIMENSIONAL SPACE-TIME 42
6-A.3 TWO-TIER QUANTUM COORDINATES FORMALISM 44
6-A.4 Y^μ GAUGE 44
6-A.5 TWO-TIER UNCERTAINTY PRINCIPLE 47
6-A.6 QUANTUM MECHANICS VS. QUANTUM FIELD THEORY VS. TWO-TIER QUANTUM THEORY 47
6-A.7 TWO-TIER PERTURBATION THEORY 48
6-A.8 TWO-TIER UNITARITY 48
6-A.8.1 Two-Tier Analyticity 49
6-A.9 TWO-TIER QUANTUM SPACE THEORY 49
6-A.10 CQ MECHANICS – A UNION OF CLASSICAL AND QUANTUM MECHANICS 49
6-A.11 PSEUDOQUANTUM FIELD THEORY 50
6-A.11.1 Color Confinement 51
6-A.11.2 PseudoQuantum Quantization for Non-Static Coordinate Systems 51
6-A.11.3 Advantages of PseudoQuantum Quantization 51
6-A.12 COMBINATION OF TWO-TIER AND PSEUDOQUANTUM FORMALISMS 52
6-A.13 A MODEL ILLUSTRATING SCALAR FIELD QUANTIZATION USING X^μ 52
6-A.14 SCALAR FEYNMAN PROPAGATORS 53

7. THE HYPERCOSMOS AND OUR N = 7 UNIVERSE 55

7.1 FUNDAMENTAL FERMION SPECTRUM 55
7.2 THE SYMMETRY GROUPS FOR N = 7 SPACE, QUEST AND UST 58
7.3 GROUPS SPECIFIED BY IRREDUCIBLE REPRESENTATIONS DIMENSIONS 58
7.4 THE N = 7 QUEST FEATURES IN THE UST 59
7.5 THE STANDARD MODEL 59

APPENDIX 7-A. GENERATION AND LAYER GROUPS 61

7-A.1 U(4) GENERATION GROUPS 61
7-A.2 U(4) LAYER GROUPS 62

7-B. HYPERCOMPLEX SPACE-TIME COORDINATES AND CONNECTION GROUP SYMMETRIES 65

7-B.1 HYPERCOMPLEX COORDINATES TRANSFORMED TO SYMMETRY GROUPS IN OUR UNIVERSE N = 7 65
7-B.1.1 The U(2) Connection Groups 68
7-B.1.1.1 Horizontal Lines 69
7-B.1.1.2 Vertical Lines 69
7-B.1.2 The Connection Groups are UltraWeak Interactions 70
7-B.2 NEWUTMOST WITH SIX REAL SPACE-TIME COORDINATES (DIMENSIONS) 70
7-B.3 NEWMAXIVERSE WITH EIGHT REAL COORDINATES (DIMENSIONS) 72

7-B.4 Determining the Connection Groups for a Space 72

8. FUNDAMENTAL REFERENCE FRAMES 75

8.1 Set of Fermions Expressed as Repeated Sets of Fermions in QUeST and UST 75
8.2 Set of Symmetry Groups Expressed as Repeated Sets of Groups in QUeST and UST 77
8.2.1 Are there Dynamics in the Fundamental Reference Frame? *78*
8.3 General Relativistic Transformation from a Fundamental Reference Frame to a Static Coordinate System 79
8.4 Contents of the Fundamental Reference Frame 80
8.5 Contents of the Higher Dimensions FRF Frame 81
8.6 Meaning of Space-Time in the HyperCosmos 83
8.7 A Physical Implementation of Platonic Realism: Plato's Theory of Abstract Forms 84

9. SPACE-TIME DETERMINATION OF PARTICLES AND SYMMETRIES 85

9.1 The Four Fermion Species Becomes Eight Subspecies for N = 7 85
9.2 Scalar Bosons of HyperCosmos Symmetries for N = 7 86
9.3 Symmetries Separation Due to Space-Time 89
9.4 N = 7 Fundamental Reference Frames Contents 89
9.5 Total View of N = 7 HyperCosmos Symmetries 89
9.6 Extension to Higher Dimension Spaces and FRF's 90

10. UNIFICATION: HYPERCOSMOS AND F-THEORY 93

10.1 The HyperCosmos Space and Unification Space of Our Universe 93
10.2 HyperCosmos Grand Unification 94
10.2.1 Form of Our Universe Dimension Array and Symmetries *98*
10.3 Internal Symmetry GUT in F-Theory 100

11. HYPERCOSMOS AND UST SYMMETRY SPLITTINGS 101

11.1 Symmetry Splitting in our N = 7 Universe 101
11.1.1 Cayley Number View *101*
11.1.2 Fundamental Representation Derivation (FRF) *101*
11.2 Symmetry Splitting in a Space N Universe 102
11.2.1 Cayley Number View of Space N *102*
11.2.2 Fundamental Representation Derivation (FRF) *102*
11.3 Scalar Bosons ElectroWeak-like Symmetry Breaking 103
11.4 The Standard Model 103
11.5 Evidence for HyperCosmos Structure and Two Time Coordinate Splitting 105

12. N = 7 UNIVERSE INTERACTIONS 107

12.1 Dynamics and Unification 107
12.2 Vector Boson and Graviton Lagrangian Terms 113

13. THE STRONG INTERACTION SECTOR AND THE HIGGS-CONFINEMENT DICHOTOMY 115

13.1 Higgs Mechanism – Quark Confinement Dichotomy 115
13.2 Quark Confinement Potential 116
13.3 Strong Interaction Lagrangian Terms 116
13.3.1 Canonical Equal Time Commutation Relations *118*
13.3.2 Dynamical Equations *119*
13.4 Strong Interaction Potential 120
13.5 The Origin of the Linear Potential 121

14. GRAVITATIONAL POTENTIAL ON THREE DISTANCE SCALES VS. THE MOND ALTERNATIVE 123

14.1 Gravitation Sector Dynamic Equations 124
14.2 Gravity Potential 126
14.3 Solution for the Case of a Massless Graviton 127
14.4 Intra-Solar System Distance Scale 128
14.5 Galactic Distance Scale 129
14.6 Intergalactic Distance Scale 129
14.7 Theoretical Agreement With Gravitational Data at all Known Distances 129

REFERENCES 131

INDEX 139

ABOUT THE AUTHOR 143

FIGURES and TABLES

Figure 1.1. The HyperCosmos spaces spectrum related to a higher space space-time dimension and its dimension array. 2
Figure 1.2. Separation of symmetries due to two time coordinates. The separation by two light speeds may account for the Standard Model form of the interaction groups. See chapter 9 for details........ 3
Figure 1.3. Block form of a 16 × 16 NEWQUeST/NEWUST dimension array with each block row corresponding to one layer. For the fermion case each block contains four generations. The result is 4 × 4 blocks and a total of 256 fermions. 4
Figure 1.4. Fermion particle spectrum and partial examples of the pattern of mass mixing of the Generation groups and of the Layer groups. Unshaded parts are the known fermions, A shaded generation is an additional, as yet not found, 4th generation. The lines on the left side (only shown for one layer) display the Generation mixing within each layer. The Generation mixing occurs within each layer using a separate Generation group for each layer. The lines on the right side show Layer group mixing (for Dark matter) with the mixing among all four layers for each of the four generations individually. There are four Layer groups for Normal matter and four Layer groups for Dark matter.. There are 256 fundamental fermions. QUeST and UST have the same fermion spectrum. Each row is a 16 fermion subblock. 5
Figure 1.5. The pattern of symmetries of UST showing the splitting into sections of 4's, 8's, 16's and 32's. Note each layer has its own set of internal symmetries. Normal and Dark sector symmetries are also different. The shaded parts are yet to be found experimentally. As shown, the pattern of symmetries closely matches the QUeST/UST dimension array splittings.The 4 ×4 blocks match the symmetry splits. SU(4) is split into SU(3)⊗U(1) experimentally.The eight SL(2, **C**) groups are transformed into one 4-dimension space-time and seven U(2) Connection Groups........ 6
Figure 1.6. HyperCosmos spectrum plotted as an Inverted energy spectrum with E = – d_{dN}. See Fig. 1.3 below. 10
Figure 2.1. A segement of the one dimension space embodied in a string of infinite length extending from -∞ to ∞. The center of force is indicated by the filled circle...... 17
Figure 2.2. The d_{dN} part of the positive energy levels of the UniDimension ProtoCosmos Model (not drawn to scale and a d_{dN} minus sign is not shown). The various quantities match the HyperCosmos spaces spectrum in Fig. 1.1. The energy levels are symbolically depicted. Negative N levels are not shown. 18
Table 3.1. Spins of fermion wave functions in HyperCosmos spaces. Higher spin wave functions are acceptable with divergence-free perturbation theories if Quantum Field Theory is formulated using the Two-Tier formalism. 21

Figure 3.1. The energy levels of a Tower ProtoCosmos. It is not drawn to scale. The various quantities are from the HyperCosmos spaces spectrum in Fig. 1.1. The energy levels are symbolically depicted on the tower. The • represents the Tower ProtoCosmos point space. 23
Figure 3.2. Amplitude vs. spin qualitative diagram for eq. 3.10 where $\alpha(s) = \sqrt{E} = 16s + 8$ where the spin is treated as a continuous parameter. Each "dot" corresponds to a HyperCosmos space. The spaces are Blaha numbered as in Fig. 1.1. They correspond to spins for $N \le 9$ appearing above in Table 3.1 based on eqs. 3.9a and 3.9b. 24
Figure 6.1. The CASe groups of the HyperCosmos spaces. 40
Figure 7.1. 16×16 dimension array. 55
Figure 7.2. The QUeST 16×16 fermion spectrum of our universe tentatively arranged as SU(4)-plets that correspond directly with SU(4) (or SU(3)⊗U(1)) fermions. There are four layers. Each set of 4 fermions has 4 generations matching the number of rows in each layer. This Periodic Table is broken into Normal and Dark sectors. 56
Figure 7.3. QUeST Fermion particle spectrum and partial examples of the pattern of mass mixing of the Generation groups and of the Layer groups. Unshaded parts are the known fermions, A shaded generation is an additional, as yet not found, 4th generation. The lines on the left side (only shown for one layer) display the Generation Group mixing within each layer. The Generation mixing occurs within each layer using a separate Generation group for each layer. The lines on the right side show Layer group mixing (for Dark matter) with the mixing among all four layers for each of the four generations individually. There are four Layer groups for Normal matter and four Layer groups for Dark matter. There are 256 fundamental fermions. QUeST and UST have the same fermion spectrum. 57
Figure 7.4. The SU(4) based pattern of symmetries of QUeST and UST. Note each layer has its own set of internal symmetries. Normal and Dark sector symmetries are also different. The Generation Groups, theLayer Groups and the Connection Groups (not shown here) tie the various parts together. The origin and role of these groups is described in Appendices 7-A and 7-B. The pattern of symmetries matches those of QUeST. SU(4) is broken to SU(3)⊗U(1) experimentally.The eight SL(2, **C**) groups are transformed into one 4-dimension space-time and seven U(2) Connection Groups in Appendix 7-B. 60
Figure 7.5. The SU(3)⊗U(1) form of the 4 layers of symmetry groups of an N = 7 universe of the form in QUeST and UST. Seven of the SL(2, **C**) groups are transformed to U(2) Connection groups later. See Appendix 7-B. 60
Figure 7-B.1. The four layers of QUeST, NEWUST and NEWQUeST internal symmetry groups (and space-time) with SU(4) before breakdown to SU(3)⊗U(1). Note the left column of blocks are combined below to specify a 4 dimension real space-time

plus seven U(2) Connection groups. Note each layer has 64 dimensions = 56 + 8 dimensions. 67
Figure 7-B.2. The seven U(2) Connection groups (shown as 10 lines) between the eight NEWQUeST/NEWUST blocks. Connection groups are obtained by transfering 28 dimensions from QUeST space-time to internal symmetries with the consequent reduction of the space-time from four octonion (complex quaternion) coordinates to four real coordinates. The Connection groups generate rotations and interactions between corresponding fermions and vector bosons of each pair of blocks. The Normal and Dark sector U(2) vertical connections above (E, F, G) represent the same U(2) groups. 68
Figure 7-B.3. NEWUTMOST has four NEWQUeST copies. An SU(4) internal symmetry Connection group maps between corresponding fermions in the four copies: fermion by fermion. An additional U(1) Connection group applies to every corresponding fermion. It is not shown in this figure. 71
Figure 7-B.4. The SU(4) Connection Group of NEWUTMOST connecting fermions in the four NEWQUeST "copies" blocks. An additional U(1) Connection group applies to every corresponding fermion. It is not shown in this figure. 71
Figure 7-B.5. The SU(8) Connection Group of NEWMaxiverse connecting fermions in the four NEWUTMOST "copies" blocks. 72
Figure 8.1. Four layers of fermions with each layer containing four generations. The fermions are arranged in SU(4)-like sets of four fermions based on the structures specified in chapter 10. The sets could be arranged to a different grouping without changing the discussion in this chapter. 76
Figure 8.2. The 16 fermions that constitute the set that is replicated 16 times in the 256 fermion spectrum for N = 7 space (universe). 77
Figure 8.3. The "initial" distribution of sets of N = 7 symmetry groups. Each set is distinct and supports interactions only for the corresponding set of fermions (separately for Normal and Dark fermions) in Fig. 8.1. *Thus each set of 16 fermion generations has its own quantum numbers and interactions.* Each U(4)⊗U(4) set has a 16 real-valued dimension representation, which will be of importance when we consider Fundamental Reference Frames in section 8.3. 78
Figure 8.4. The transformed/broken sets of symmetries in QUeST (UST) and in N = 7 HyperCosmos space.. Note each element has a 16 real dimension representation. This depiction is also evident in QUeST and the UST. The SL(2, **C**) representation has four coordinates. Seven of these groups are transformed to U(2) Connection groups. 78
Figure 8.5. The contents of the ten types of Fundamental Reference Frames for the ten spaces. The elements are fermions or symmetry fundamental representation dimensions or dimensions. 81
Figure 8.6. Possible contents of five of the ten types of Fundamental Reference Frames for fermions. The primes distinguish different fermions. 82

Figure 8.7. Possible contents of five of the ten types of Fundamental Reference Frames for symmetry dimensions. The groups may undergo a further breakdown after being mapped to a static space-time to give terms like those in Fig. 8.4. 83
Figure 9.1. Separation of fermion species due to su(1, 1) and its two times. The total number of corresponding fermions is 16 if account is taken of quarks appearing as triplets. The predicted occurrence of quark in triplets in the HyperCosmos is motivated by the need for fermions to appear in multiples of four units. Thus the separation by two light speeds may account for the Standard Model form of the fermion species and also for triplets of quarks. Experimentally, superluminal neutrinos appear to move at the speed of light due to their negligible masses. Down-type quarks, being confined within particles, do not directly display their superluminal nature. Dark matter, being unobserved may well be superluminal. Their superluminal nature may be part of the cause for their current unobservability. It is also possible that the superluminal nature of these quantities may "wash out" when a transformation is made from an FRF to a static reference frame. 86
Figure 9.2a. Real parts of bosons separated by su(1, 1)'s two times. Each set of two parts will combine, as real and imaginary parts, into a scalar boson in a General Relativistic – Internal Symmetry transformation. 87
Figure 9.2b. The sets of scalar bosons generated by the transformation in the four layers. Line (I) is for the Standard Model sector enhanced to four generations and a Dark sector. Line (II) corresponds to the Generation group and an U(2) Connection Group described earlier. Line (III) corresponds to the Layer Group also described earlier. 88
Figure 9.3. The real parts of dimensions in the N = 7 FRF's symmetry part. The separation of symmetries is due to su(1, 1) and its two times. The separation by two light speeds accounts for the UST (and Standard Model) form of the interaction groups. 89
Figure 9.4. The contents of the N = 7 symmetries. The separation of symmetries due to su(1, 1) with two times. The separation by two light speeds accounts for the QUeST/UST (and the Standard Model) form of the interaction groups. Line (I) is for the Standard Model sector enhanced to four generations and to a Dark sector. Line. (II) corresponds to the Generation group, and an U(2) Connection Group described earlier except line (II) of layer 1 which has a space-time group SL(2, **C**). Line (III) corresponds to the Layer Group also described earlier. 91
Figure 10.1. The HyperCosmos spaces spectrum related to a unification higher space space-time dimension and its dimension array. Note the spaces with $r^{\prime} > 18$ are outside the HyperCosmos set of 10 spaces. However they have the same form as the 10 HyperCosmos spaces. 95

Figure 10.2. Diagram of the relation between a space and its "companion" higher space-time dimension space with Fundamental Reference Frames indicated. The higher space sets d_{dN} for the lower space using 'Higher' space General Relativity. Fundamental Reference Frames are like the rest frames of particles. They reduce the number of elements to a core set of elements. The elements are fermions or gauge vector symmetry fundamental representation dimensions. 97
Figure 10.3 The steps for determining a dimension array using a General Relativistic transformation in the "partner" higher space. Note the numerical values generated by the higher space General Relativistic transfomation are only important in so far as they are zero or non-zero. 97
Figure 10.4. The Blaha number N = 6 dimension array quadrisected three times to N = 9 dimension arrays. Quadrisection continues down to N = 9. There are 64 N = 9 space dimension array parts visible above. The horizontal and vertical patterning are the same. This partitioning applies to both fermions and symmetries. 98
Figure 10.5. In QUeST/UST (N = 7) there are four layers of fundamental fermions with each layer containing four generations. The fermions are arranged in SU(4)-like (possibly broken to SU(3)⊗U(1)) sets of four fermions based on the structures specified in chapter 10 of Blaha (2022c). 99
Figure 10.6. The sets of symmetries in QUeST/UST and in its N = 7 HyperCosmos space. Note each boxed element has a 32 real dimension representation. This depiction is also evident in QUeST and the UST. The SL(2, **C**) representation has four coordinates. 100
Figure 11.1. The Blaha number N = 6 dimension array quadrisected three times to N = 9 dimension arrays. There are 64 N = 9 space parts visible. 104
Figure 11.2. The contents of the N = 7 symmetries with ElectroWeak-like SU(2)⊗U(1) broken groups shaded. The separation of symmetries due to su(1, 1) with two times. The separation by two light speeds accounts for the QUeST/UST (and the Standard Model) form of the interaction groups. Line (I) is for the Standard Model sector enhanced to four generations and to a Dark sector. Line. (II) corresponds to the Generation group, and an U(2) Connection Group described earlier except line (II) of layer 1 which has a space-time group SL(2, **C**). Line (III) corresponds to the Layer Group also described earlier. 105
Figure 12.1. The contents of the N = 7 symmetries with ElectroWeak-like SU(2)⊗U(1) broken groups shaded from Fig. 11.2. Line (I) is for the Standard Model sector enhanced to four generations and to a Dark sector. Line. (II) corresponds to the Generation group, and an U(2) Connection Group described earlier except line (II) of layer 1 which has a space-time group SL(2, **C**). Line (III) corresponds to the Layer Group also described earlier. 108

INTRODUCTION

The Cosmic Panorama derives the form of the Cosmos in a series of stages from the ProtoCosmos through the HyperCosmos to the Unified SuperStandard Theory (UST). As it proceeds it shows significant experimental support for all stages giving it an unsuspected reality that explains previously inexplicable features down to The Standard Model level. It explains the division of symmetries into SU(3) and ElectroWeak in the Standard Model as well as the form of the fundamental fermion spectrum. It explains the Dichotomy between the Higgs Mechanism and Strong Interaction Confinement as a manifestation of the two time coordinates of the HyperCosmos unification space and F-Theory. The separation of internal symmetries in the UST is shown to be a structural feature of HyperCosmos spaces – an intimate connection between HyperCosmos structure and symmetries. The result is a new view of the origin of the separation of symmetries as mostly structural – not due to symmetry breakdown. Only the ElectroWeak-like symmetries enjoy a Higgs breakdown.

Other noteworthy features include: an improved UniDimension ProtoCosmos, a Tower ProtoCosmos extension, a Regge Trajectory Resonances form for HyperCosmos spaces, a new form of ProtoCosmos wave functions, an improved unification in 12 dimensions for our universe with F-Theory (a SuperString theory), the introduction of the Higgs-Confinement Dichotomy that shows the Higgs Mechanism and the Strong Interaction Confinement Interaction to emanate from the two time coordinates of su(1,1) and 12 dimension space, and a new specification of the Riemann-Christoffel curvature tensor for UST including a higher derivative SU(3) interaction giving quark confinement and a related formulation of higher derivative Gravity with implications similar to MOND.

This book explores all known features of the HyperCosmos and the UnifiedSuperStandard Theory.

1. The HyperCosmos and Unified SuperStandard Theory (UST)

1.1 Synopsis of the HyperCosmos and the Unified SuperStandard Theory

Three years ago this author found that the Unified SuperStandard Theory, which was developed over a period of years preceding 2019, had a fundamental fermion spectrum and a set of symmetries very similar to a universe theory based on hypercomplex coordinates. Subsequently this author developed, and refined, a hypercomplex space theory for our universe, and a general theory of ten hypercomplex spaces that could support other universes of different space-time dimensions such as the Megaverse (Multiverse). This theory became the HyperCosmos theory.

The HyperCosmos theory has evolved to provide a description of the structure of our universe and other possible universes. We were able to base the HyperCosmos theory on a more fundamental theory called the UniDimension ProtoCosmos theory,[1] which had a wider spectrum of spaces. It included the author's previously considered ten spaces of the HyperCosmos[2] and added an infinity of other spaces of fractional dimension generated as an accumulation point of the UniDimension ProtoCosmos spectrum of spaces. We called this additional set of spaces the Limos[3] (after the Lilliputians of *Gulliver's Travels*.)

Equally importantly, we found that one could base the spectrum of fundamental particles and symmetries of each space on a much smaller spectrum of particles and symmetries in a Fundamental Reference Frame (FRF).[4] The spectrum of particles and symmetries in each FRF replicate to form the full spectrums in each HyperCosmos space. The replication is implemented by a General Relativistic transformation[5] in a higher space partner of each space. See Fig. 1.1 below. For example, the space of our universe has a 12 dimension higher unification space. Remarkably the unification of the symmetries of our space is implemented in the 12 dimension higher space. This higher space has the overall features of a SuperString F-Theory of 12 dimensions with two time dimensions.

As a result we have unification for any universe of any of the ten HyperCosmos spaces. Each universe of each space type is endowed with a higher dimension unification space that generates its dimension array as a Grand Unified Theory (GUT),

---

[1] See Blaha (2022e).

[2] A space defines a space-time and internal symmetry dimensions. A universe is an instance of a space that is created with an initial energy. Each universe has a dimension array containing the fundamental representation dimensions of its internal symmetries and space-time.

[3] Blaha (2022d).

[4] Blaha (2022c).

[5] The General Relativistic transformations are implemented within the author's GiFT formulation of Quantum Field Theory. GiFT combines the author's PseudoQuantum and Two-Tier formulations. See Blaha (2021j).

and a set of General Relativistic transformations providing Unification. Each universe also has a FRF that reduces its dimension array to a much smaller number of dimensions and thus a much smaller number of fundamental fermions and a much smaller number of symmetries. An analysis of these smaller sets of symmetries and dimensions reduces them to constituent units whose structure is determined by two time coordinates' sublight and superluminal splittings in the space's associated higher space.[6] See chapter 9 and Fig. 1.2 below. Our universes' space has a 12 dimension space-time with two time coordinates—equivalent to F-Theory space.

THE HYPERCOSMOS SPACES SPECTRUM

Blaha Space Number $N = O_s$	Cayley-Dickson Number n	Cayley Number d_c	Dimension Array column length d_{cd}	Dimension Array Size d_{dN}	Space-time-Dimension r	Higher Space-time r' Source of d_{dN} r'	Higher Space Array d_{dN}'
0	10	1024	2048	2048^2	18	40	2^{44}
1	9	512	1024	1024^2	16	36	2^{40}
2	8	256	512	512^2	14	32	2^{36}
3	7	128	256	256^2	12	28	2^{32}
4	6	64	128	128^2	10	24	2^{28}
5	5	32	64	64^2	8	20	2^{24}
6	4	16	32	32^2	6	16	2^{20}
7	**3**	**8**	**16**	$\mathbf{16^2}$	**4**	**12**	$\mathbf{2^{16}}$
8	2	4	8	8^2	2	8	2^{12}
9	1	2	4	4^2	0	4	2^{8}

Figure 1.1. The HyperCosmos spaces spectrum related to a higher space space-time dimension and its dimension array.

1.1.1 Unified SuperStandard Theory

The hypercomplex theory that ultimately emerged began as an attempt by this author in autumn, 2019 to find a deeper basis for his Unified SuperStandard Theory (UST). It became clear that a hypercomplex theory could be the basis of UST. Since then the author developed Octonion Cosmology based on a study of fermion wave function spin in higher space-time dimensions.[7] In the autumn, 2021 the author found that a consideration of General Relativistic transformations of fermion creation and annihilation operators in PseudoQuantum Field Theory led to a much simpler, much improved spectrum of spaces. See Fig. 1.1. This effort was the origin of HyperCosmos Theory[8] with a clear relation to fermion creation/annihilation operators. Since then the author has considered the possibility of fractional particles with a lattice-like structuring

---

[6] Blaha (2022c) and earlier books.
[7] Blaha (2020x).
[8] Blaha (2021i) through (2022c).

of fundamental particle interiors, and also connections to partons and Dual Resonance Models (Veneziano amplitudes).

The UST has a dimension array that maps to a fermion spectrum and to a symmetry groups spectrum.[9] The maps are well-defined by the hypercomplex structuring of the set of HyperCosmos spaces. Figs. 1.3, 1.4 and 1.5 are for our universe and appear in QUeST which corresponds to the Blaha number N = 7 HyperCosmos space. They separate fermions and symmetries in such a way as to raise the question of the role of symmetry breaking in the Standard Model. The structuring is not dependent on coupling constants and gauge vector boson masses. See chapters 4 and 5 for details related to Figs. 1.3 – 1.5.

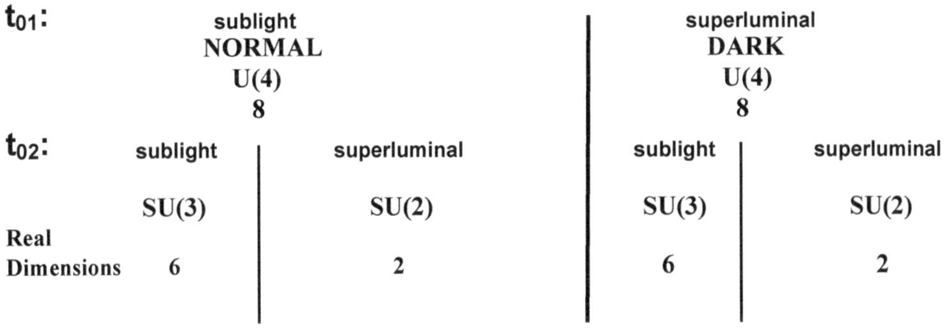

Figure 1.2. Separation of symmetries due to two time coordinates. The separation by two light speeds may account for the Standard Model form of the interaction groups. See chapter 9 for details.

[9] The Dark fermions and symmetries are placed on an equal footing in the UST. The Dark sector could be eliminated if the HyperCosmos set of spaces was redefined.

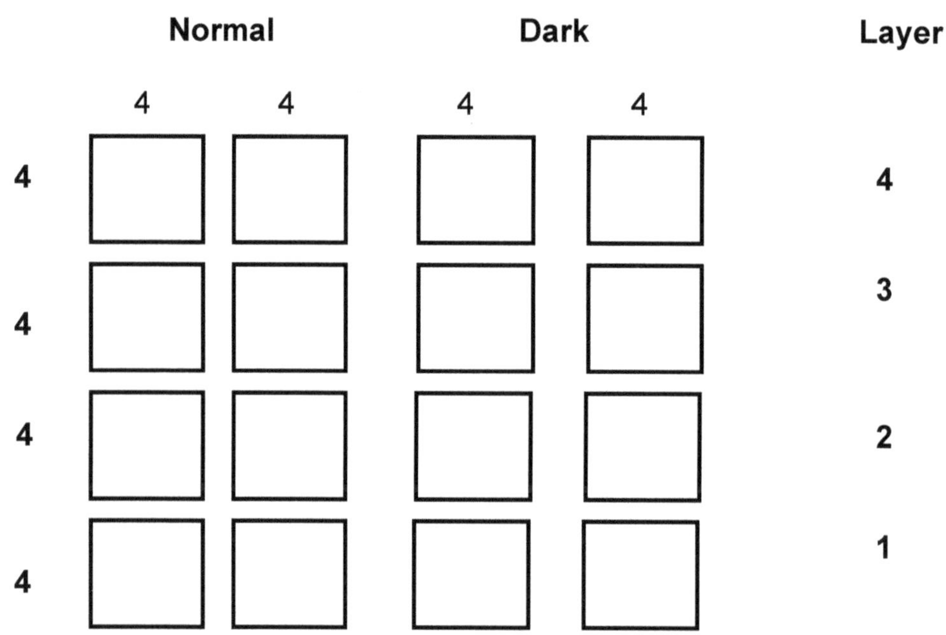

Figure 1.3. Block form of a 16 × 16 NEWQUeST/NEWUST dimension array with each block row corresponding to one layer. For the fermion case each block contains four generations. The result is 4 × 4 blocks and a total of 256 fermions.

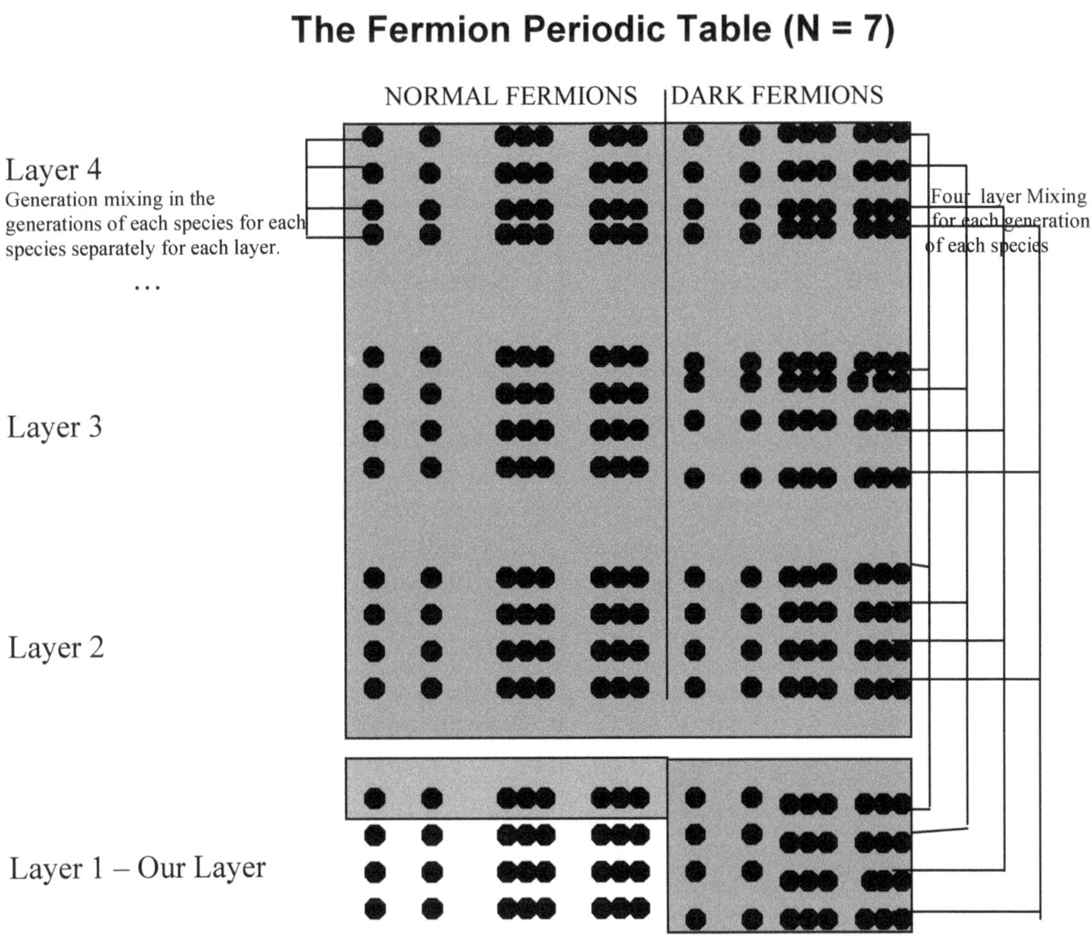

Figure 1.4. Fermion particle spectrum and partial examples of the pattern of mass mixing of the Generation groups and of the Layer groups. Unshaded parts are the known fermions, A shaded generation is an additional, as yet not found, 4th generation. The lines on the left side (only shown for one layer) display the Generation mixing within each layer. The Generation mixing occurs within each layer using a separate Generation group for each layer. The lines on the right side show Layer group mixing (for Dark matter) with the mixing among all four layers for each of the four generations individually. There are four Layer groups for Normal matter and four Layer groups for Dark matter.. There are 256 fundamental fermions. QUeST and UST have the same fermion spectrum. Each row is a 16 fermion subblock.

NEWQUeST/NEWUST Symmetries Table

Number of Columns =	4 NORMAL	4	4 DARK	4
Layer 1 4	SU(4)⊗ Generation U(4)	SU(2)⊗U(1)⊗SL(2,C)[10] Layer U(4)	SU(4)⊗ Generation U(4)	SU(2)⊗U(1)⊗SL(2,C) Layer U(4)
Layer 2 4	SU(4)⊗ Generation U(4)	SU(2)⊗U(1)⊗SL(2,C) Layer U(4)	SU(4)⊗ Generation U(4)	SU(2)⊗U(1)⊗SL(2,C) Layer U(4)
Layer 3 4	SU(4)⊗ Generation U(4)	SU(2)⊗U(1)⊗SL(2,C) Layer U(4)	SU(4)⊗ Generation U(4)	SU(2)⊗U(1)⊗SL(2,C) Layer U(4)
Layer 4 4	SU(4)⊗ Generation U(4)	SU(2)⊗U(1)⊗SL(2,C) Layer U(4)	SU(4)⊗ Generation U(4)	SU(2)⊗U(1)⊗SL(2,C) Layer U(4)

Figure 1.5. The pattern of symmetries of UST showing the splitting into sections of 4's, 8's, 16's and 32's. Note each layer has its own set of internal symmetries. Normal and Dark sector symmetries are also different. The shaded parts are yet to be found experimentally. As shown, the pattern of symmetries closely matches the QUeST/UST dimension array splittings.The 4 ×4 blocks match the symmetry splits. SU(4) is split into SU(3)⊗U(1) experimentally.The eight SL(2, **C**) groups are transformed into one 4-dimension space-time and seven U(2) Connection Groups.

---

[10] The Lorentz group. $SO^{+}(1,3)$ in particular

1.1.2 The Standard Model Theory

The structure of the Standard Model of Elementary Particles of our universe is present in the Unified SuperStandard Theory. The known fermions appear as unshaded in Fig. 1.4. The known symmetries appear as unshaded in Fig. 1.5. The unknown fermions seem to be of large mass beyond the reach of current accelerators. The gauge vector bosons of the unknown symmetries also appear to be of large mass.

The Dark sectors were placed in the UST as symmetrical to the normal sectors. Their discovery remains to be done. If the Dark sectors are absent then a revision of the HyperCosmos set of spaces would be necessary. Such a revision would be straightforward and preserve the general characteristics of the HyperCosmos.

The universes of the other higher HyperCosmos spaces have their own USTs, which contain multiple replicates of our UST. For example the next higher space (N = 6) has a Multiverse (Megaverse) instance. It has four replicates of our UST.

For each space above our space, the universe instance has quadruple replicates of the fermions and symmetries of the space/universe immediately below it.

1.1.3 Limos and Fractionation

The Limos spaces within the HyperCosmos set of spaces support the fractionation of particles and symmetries as described in Blaha (2022b). Fractionation is analogous to the decomposition of space-time in lattice gauge theory. Fractionation is a more comprehensive decomposition. Particles and symmetries, as well as space-time, are fractionated. In the limit of infinite decomposition quantum theoretic aspects become classical and symmetries become abelian. In this limit a model of the interior "space" of particles using gambols may be constructed.

Gambols appear to be a theoretical basis of partons in Deep Inelastic lepton-nucleon scattering. Also a gambol view of harmonic oscillator models of Dual Resonance phenomena is supported leading to the Veneziano Amplitude.

1.2 The Logic of the Derivation of the Unified SuperStandard Theory

The logic of the derivation has several stages. We will specify the stages here and then provide the details in the chapters that follow.

1.2.1 The ProtoCosmos Stage

This stage defines a UniDimension ProtoCosmos[11] of one dimension. It has a one dimension Dirac-like equation that generates a bound state spectrum using a one dimension electromagnetic-like potential. The Dirac-like wave function has a factor that is a function of the one dimension coordinate and a factor that is a function of the set of coordinates of an embedded universe. We call the wave function a space wave function since one factor embodies a space. (See Chapter 5.)

The bound state spectrum had the geometric form:

$$E_N \cong m - d_{dN} \tag{1.1}$$

where m is the mass and where

[11] Blaha (2022e).

$$d_{dN} = 2^{22-2N} = 2^{r+4} \tag{1.2}$$

d_{dN} is the number of components in the dimension array of Blaha number N. Thus the spectrum of HyperCosmos spaces of Fig. 1.1 emerges. Note the bound state accumulation point gives the Limos spaces.

We use the relation between Dirac matrix column length γ_C, spin s, and space-time dimension r:

$$\gamma_C = 2^{r/2} = 4s + 2 \tag{1.3}$$

implying

$$r = \ln_2(4s + 2)^2$$

where r is the number of space-time dimensions. Then we find the energy E_N satisfies

$$\begin{aligned} E_N = m - d_{dN} &= m - 2\wedge(\ln_2(4s + 2)^2 + 4) \\ &= m - 16(4s + 2)^2 \end{aligned} \tag{1.4}$$

or

$$E' = \sqrt{(m - E_N)} = 8 + 16s \tag{1.5}$$

is linearly related to the spin of the universe factor of the wave function. This relation is reminiscent of Regge pole trajectories. It leads to the speculation that the higher spaces of the HyperCosmos may be viewed as "resonances" (excitations) of lower spaces. (See chapter 4.) The ProtoCosmos generates the HyperCosmos spectrum of spaces in the UniDimension Model and the Tower Model.

1.2.1.1 Generation of Spaces from an Energy

The energy relation implied by eqs. 1.1 and 1.2 for m = 0 and E'' = -E is

$$r = \ln_2 E'' - 4 = \ln_2 E''/16 \tag{1.6}$$

If we require r to be an integer then E'' must have the form

$$E'' = 2^k = 2^{22-2N} \tag{1.7}$$

for k = 4, 5, 6, … or N = 0, 1, 2, … 9. Thus we obtain part of the HyperCosmos spectrum of spaces. If we allow k < 4 values that give r values that are fractional powers of 2 then we obtain the Limos part of the HyperCosmos spectrum.

Thus one may envision a zero space-time dimension ProtoCosmos where a point (the Tower ProtoCosmos) has a series of allowed values for E'' and a set of spaces is generated of the form of the HyperCosmos spectrum. Chapter 4 shows each space has an energy E'' and spin s satisfying

$$E'' = 16(4s + 2)^2 \tag{1.8}$$

as seen above. We define a ProtoCosmos amplitude α(s) with

$$\alpha(s) = \sqrt{E''} = 4(4s + 2) \tag{1.8a}$$

α(s) has a linear dependence on spin s similar to a Regge pole trajectory. At certain spins we find corresponding HyperCosmos spaces. We specify them as functions of Blaha number N:

$$\alpha(s_N) = \sqrt{E''_N} = 4(4s_N + 2) \tag{1.8b}$$

in analogy with Regge pole trajectory resonances. See chapter 2 for details.

Since the dimension array column length satisfies

$$d_{cdN} = \sqrt{d_{dN}} = \sqrt{E''_N} = \alpha(s_N) \tag{1.9}$$

we find

$$d_{cdN} = 16s_N + 8 \tag{1.10}$$

where s_N is the Blaha space N wave function spin.

The ProtoCosmos then has a spaces trajectory similar to Regge pole trajectories. If we allow N to be continuous then we obtain a continuous spectrum of space-time dimensions r and non-integer-based HyperCosmos spaces interspersed within the integer-based spaces in Fig. 1.1.

These considerations raise the possibility of a deeper synthesis of the ProtoCosmos. See chapter 2.

1.2.2 The HyperCosmos Stage

The spectrum of spaces that we found for the HyperCosmos invites the view that it corresponds to an inverted spectrum of states where d_{cd} (of Fig. 1.2) plays a role analogous to (negative) energy, and the Blaha space number N = O_s plays the role of an energy level index. First we look at the inverted HyperCosmos spaces spectrum in Fig 1.6 below. It reveals a Hydrogen-like energy spectrum. Note the energy accumulation point at zero as N = $O_s \to \infty$. Note also the geometric scaling behavior of the energy d_{cd}. Fig. 1.6 shows the inverted spectrum plotted as an energy spectrum.

Inverting the energy scale (and making the energy sign positive) gives the HyperCosmos space spectrum. Chapters 2 and 3 developed the ProtoCosmos. Chspter 4 shows it has an amplitude – spin relation that gives the space spectrum a Regge trajectory analogue. Chapter 5 describes the wave function solutions of the Unidimension ProtoCosmos that each contains a different universe part.

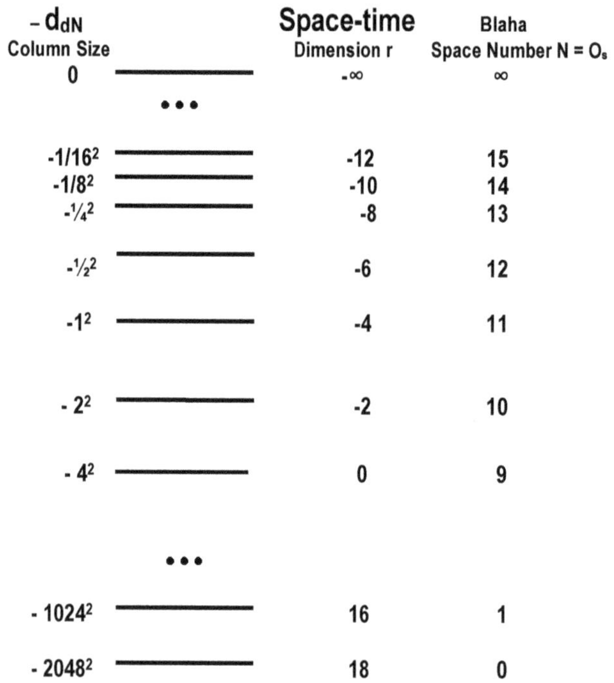

Figure 1.6. HyperCosmos spectrum plotted as an Inverted energy spectrum with $E = - d_{dN}$. See Fig. 1.3 below.

1.2.3 The QUeST Stage

Our universe is based on the N = 7 HyperCosmos space. QUeST was its original Model.

1.2.4 The UST Stage

QUeST originated in the Unified SuperStandard Theory (UST) developed by the author a few years ago. The author found QUeST and UST had almost the same set of features.

1.2.5 The Standard Model Stage

The UST originated as a generalization of the Standard Model. It added Two-Tier quantization to avoid divergences in perturbation theory. Two-Tier theory also enables higher dimension quantum field theories to avoid divergences.

The UST also embodied the author's PseudoQuantum Theory develop by the author in the 1970s. PseudoQuantum Theory led to the HyperCosmos and its use of Fundamental Reference Frames for unification of General Relativity and Internal Symmetries.

Appendix 1-A. Experimental Support for the HyperCosmos and UST

There is evidence for the ProtoCosmos and the HyperCosmos. It is not direct evidence since we are bound by our universe and its elapsed lifetime. It is evidence that indirectly supports their existence and their structure.

1. Direct Generation of the HyperCosmos spectrum from a UniDimension ProtoCosmos.

2. Relic Structure in Internal Symmetries and the Fundamental Fermion Spectrum. The Standard Model internal symmetry SU(2)⊗U(1)⊗SU(3) is shown to appear in HyperCosmos/QUeST as a natural separation not requiring symmetry breaking Higgs fields. QUeST introduces additional symmetries that add to those of the Standard Model in a direct way—again without Higgs bosons. Higgs breaking is restricted to ElectroWeak-like sectors.

3. *The separation of fermions and symmetries into superluminal and subluminal parts in the Fundamental Reference Frame leads to the fundamental fermion spectrum, to the set of internal symmetries, and to the understanding of the relation of the Higgs Mechanism and Strong Interaction Confinement (the Higgs-Confinement Dichotomy). See chapter 9.*

4. The Dual Resonance Model of hadron scattering is *directly* based on the HyperCosmos using fractionated hadrons. This approach is an alternative to String Theories. This approach supports hadronic Veneziano amplitudes.

5. The Parton Model of Deep Inelastic Lepton-Nucleon Scattering may be viewed as based on fractionated particles. One finds that it "explains" quasi-free partons, the apparent lack of strong interactions between partons thus giving parton freedom, and scaling in x, as well as other features.

6. The fractionated particle formalism has lattice theory as a restricted special case.

7. The HyperCosmos leads to a deep unification based on its Fundamental Reference Frame formalism. For example, in the case of our universe, there is a unified 12 space-time dimension formalism that generates the unified General Relativity and Internal Symmetries of our universe.

2. UniDimension ProtoCosmos Model

Pascal: Man/woman is a thinking reed.
Physics: The Cosmos is a blossoming tree.

2.1 New UniDimension ProtoCosmos Model

The ProtoCosmos UniDimension Model consists of a one dimension "line" having a center of force. (Fig. 2.1) It has a one dimension (relativistic) Dirac equation with a set of space wave eigenfunctions.[12] These space wave functions have an energy spectrum that maps to the HyperCosmos spectrum of spaces.

The UniDimension ProtoCosmos Model is the most minimal formulation of the basis of the HyperCosmos.

2.1.1 UniDimension Force

The UniDimension force in the ProtoCosmos will be treated as a one dimension electromagnetic potential. It consists of a one potential A(x) wave function only due to the ProtoCosmos having a one dimension space with no time coordinate. We view this space as the limit of a two dimension space with a time component where the time component is "frozen."

We assume the electromagnetic interaction has a 1/x form knowing that a geometric scaling output energy spectrum will result that leads to the HyperCosmos. It is generated from a Lagrangian term:

$$\mathcal{L} = -\tfrac{1}{2} m_A^2 A^2 \tag{2.1}$$

where m_A is a mass-like parameter corresponding to A, which is massless. The dynamic equation for A is[13]

$$m_A^2 A = 0 \tag{2.2}$$

If we restrict x to $0 \le x \le \infty$, and use

$$\int_0^\infty dk\, e^{ikx} = i/(x + i\varepsilon)$$

then

$$A = V(x) = 1/(2\pi\, m_A^2) \int_0^\infty dk\, e^{-ikx} = 1/(\, m_A^2 x) \tag{2.3}$$

The electromagnetic propagator has the form

---

[12] Space wave functions are wave functions that have a set of coordinates of the current universe and a set of coordinates for a "subsidiary" universe. Seehapter 5 here and cjchapter 6 in Blaha (2022c) for details.

[13] Note gauge invariance is absent.

$$S(x - y) = \langle 0|T(A(x)A(y))|0\rangle = 1/(2\pi) \int dk\, e^{-ik\cdot(x-y)}/(m_A{}^2) \tag{2.4}$$

Note

$$m_A{}^2\, S\,(x - y) = \delta(x - y) \tag{2.5}$$

The resulting basic UniDimension Dirac-like equation is:

$$(p + m + V(x))\psi = H\psi = E\psi \tag{2.6}$$

where m is a fermion mass, V specifies a force center at x = 0, and

$$p = \hbar/i\; d/dx \tag{2.7}$$

$$V(x) = -Z/x \tag{2.8}$$

for a central charge Z.[14] Below we extend eq. 2.6 to space wave functions incorporating subsidiary universes. Note the Dirac matrix in one dimension is a 1 × 1 constant that we can take to be unity.

2.1.2 ProtoCosmos Free Space Wave Function

We wish to define a wave function for the UniDimension that embodies both the x line and introduces a subsidiary space, from which the HyperCosmos spaces emerge. We generalize eq. 2.6 to the separable equation

$$[p + m + V(x) + M + i\gamma^{\mu}\cdot\partial/\partial u^{\mu}|_{f(E)}]\psi(x, u) = H\psi(x, u) = E\psi(x, u) \tag{2.9}$$

where the universe u is described by the fermion wave, where x is the ProtoCosmos coordinate and u represents the subsidiary space/universe coordinates. The set of u coordinates consists of $(u_1, u_2, \ldots, u_{f(E)})$ where f(E) is an integer specified as a function of the energy E. We separate eq. 2.9 into

$$\psi(x, u) = \psi_1(x)\psi_2(u) \tag{2.10}$$

$$[p + m + V(x)]\psi_1(x) = E\psi_1(x) \tag{2.11}$$

$$[\, i\gamma^{\mu}\cdot\partial/\partial u^{\mu}|_{f(E)} + M]\psi_2(u) = 0 \tag{2.12}$$

with the (discrete) E spectrum determined by the ProtoCosmos space x, and f(E), which will be an integer valued function of E, determined for each E separately for each energy value in the spectrum. The energy spectrum of E values will be seen to be geometric. The result will be a set of spaces/universes that correspond to the

[14] We absorb an $m_A{}^2$ factor into Z.

HyperCosmos set of spaces. The use of f(E) is analogous to the effects of the role of eigenvalues of L^2 (angular momentum) in the Quantum Mechanics Hydrogen atom. The Hydrogen energy E eigenvalues limit the L^2 eigenvalues $l(l + 1)$ in the Hydrogen spectrum.

Case calculated quantum mechanical wave functions for singular potentials in four dimensions. His relativistic four dimension Dirac equation[15]

$$(\beta\gamma^1 p + \beta m + V(x))\psi = H\psi = E\psi \tag{2.13}$$

has the same form as eq. 2.11 above. He calculated the singular potential energy levels based on requiring a fixed phase for wave functions at the origin. This approach guarantees the required orthogonal eigensolutions.

Using Case's method eq. 2.11 leads to an energy spectrum which we may express in the form:

$$E_N \cong m[1 - c_1 c_2^{-2N}] \tag{2.14}$$

where c_1 and c_2 are constants.

We now connect tis energy spectrum to the HyperCosmos spaces spectrum by requiring

$$\begin{aligned} c_1 &= 1/m \\ c_2 &= 2^{22} \end{aligned} \tag{2.15}$$

Then the infinite UniDimension line with the form of Fig. 2.2 has

$$E_N \cong m - d_{dN} \tag{2.16}$$

where

$$d_{dN} = 2^{22 - 2N}$$

with d_{dN} being the number of components in the dimension array of the Blaha number N space. We note the following HyperCosmos relations, which hold here:

$$\begin{aligned} r &= 18 - 2N = \text{number of space-time dimensions} \\ d_{dN} &= 2^{22 - 2N} = 2^{r+4} = \text{number of dimension array components} \\ d_{cd} &= 2^{11 - N} = \text{column length of the } d_{dN} \text{ array} \end{aligned} \tag{2.17}$$

Fig. 2.2 shows the energy spectrum superimposed on the string symbolically. Fig. 1.1 shows the HyperCosmos spaces spectrum. Below we display the wave functions $\psi_1(x)$ and $\psi_2(u)$. The $\psi_2(u)$ factor is a free wave function that has its form determined by the energy eigenvalues of $\psi_1(x)$.

We now modify eqs. 2.9 and 2.12 to a form explicitly dependent on the energy.

---

[15] From Blaha (2022c).

$$[p + m + V(x) + M + i\gamma^{\mu}\cdot\partial/\partial u^{\mu}|_{r = f(E)}]\psi(x, u) = H\psi(x, u) = E\psi(x, u) \tag{2.18}$$

$$[\, i\gamma^{\mu}\cdot\partial/\partial u^{\mu}|_{r = f(E)} + M]\psi_2(u) = 0 \tag{2.19}$$

by making the space-time dimension dependent on energy with[16]

$$\begin{aligned} f(E) = r &= \ln_2[(m - E)/16] \\ &= 18 - 2N \end{aligned} \tag{2.20}$$

by eq. 2.17. Thus

$$\gamma^{\mu}\cdot\partial/\partial u^{\mu}|_{r = f(E)} = \sum_{\mu = 1}^{r} \gamma^{\mu}\cdot\partial/\partial u^{\mu} \tag{2.21}$$

Each energy E_N has a different associated space-time dimension r.[17] Fig. 1.1 shows the dependence of the universe space-time dimension on the energy eigenvalue E_N.

The expression in eq. 2.21 represents an energy dependent term that is analogous to the derivative expression for the angular momentum term L^2 in the Quantum Mechanics hydrogen atom. This term becomes l (l + 1) where l is restricted by the hydrogen atom energy. Eq. 2.21 may be transformed to a term analogous to L^2 in a deeper more complete UniDimension ProtoCosmos that may eventually emerge.

The solutions of eq. 2.18 (in the PseudoQuantum form) are:[18]

$$\psi_{iMN\alpha}(\mathbf{x}, t, \mathbf{u}, t_u) = \psi_{1iN}(x)\, \psi_{2iMN\alpha}(u)$$

where

$$\psi_{iMN\alpha}(\mathbf{x}, t, \mathbf{u}, t_u) = \psi_{1iN}(x) \Sigma_s \int d^{r-1}q\, \mathfrak{N}(q)[u(q, s)b_{iMN\alpha}(q,s)e^{-iq\cdot u} + v(q, s)d_{iMN\alpha}(q,s)^{\dagger}e^{iq\cdot u}] \tag{2.22}$$

$$\psi_{2iMN\alpha}(\mathbf{u}, t_u) = \Sigma_s \int d^{r-1}q\, \mathfrak{N}(q)[u(q, s)b_{iMN\alpha}(q,s)e^{-iq\cdot u} + v(q, s)d_{iMN\alpha}(q,s)^{\dagger}e^{iq\cdot u}] \tag{2.22a}$$

for i = 1, 2 where $\mathfrak{N}$ is a normalization constant, where m is the mass in the ProtoCosmos space and M is the mass-energy in the universe space of Blaha number N, r is the universe space-time dimension, and α ranges from 1 through d_{dN}. We assume $q^2 = M^2$ in the universe space. Note $u = (u^0, \mathbf{u})$, and $t_u = u^0$.

The solution eq. 2.22 gives a space fermion of the type described in Blaha (2022c). The range of the symmetry index α from 1 through d_{dN} gives a dimension array of d_{dN} elements that can be subdivided ito a d_{cd} by d_{cd} square array of the type seen in

---

16 The factor of 16 figures prominently in the Fundamental Reference Frame discussions as setting the size of the basic unit of dimensions. See chapters 9 and 10, and particularly section 10.3, of Blaha (2022c) for details.

17 Note that the Hydrogen atom Schrödinger equation has a different energy dependent range of angular momentums—an analogous feature.

18 We separate the differential equation to obtain the solution in eq. 2.22. In chapter 6 of Blaha (2022c) we form a combined solution for the universe part and the internal universe parts. Physically, the solution eq. 2.22 reflects the separation of the one dimension ProtoCosmos from the subsidiary universe.

the HyperCosmos spaces spectrum. Universe states can be created from sums over α of the wave functions $\psi_{iMN\alpha}(\mathbf{x}, t, \mathbf{u}, t_u)$.

The Spaces of the UniDimension ProtoCosmos

This ProtoCosmos Model has two relevant types of space-times. First there is the one dimension ProtoCosmos "space-time." There is also the space-times of each of the model's eigensolutions. For each energy E_N there is the r space-time dimension wave function of eq. 2.22. There is also an internal r space-time dimension set of coordinates within the dimension array d_{dN}. The number of these internal space-time coordinates is the value r = 18 – 2N. Also the f(E) condition sets r for the wave function.

The internal number of space-time dimensions always equals the number of wave function space-time dimensions due to the f(E) condition. Therefore we are allowed to identify the wave function space-time with the internal space-time of each eigensolution. For each eigensolution there is one dimension ProtoCosmos space and one r dimension internal space-time. The f(E) condition causes the wave function and internal space-times of the model to be the same space-time.

Each universe of each HyperCosmos space has a wave function of the form of eq. 2.22a that exists *within* the universe. Consequently it is possible to second quantize a universe consistently within it. This may be viewed as remedying previous attempts to quantize a universe from "outside" the universe.

The eigenstates of the UniDimension ProtoCosmos correspond to the 10 universe spaces of the HyperCosmos as well as the fractional spaces of the Limos. (Fig. 1.1) The positive energy ProtoCosmos spectrum is limited by 0 (an accumulation point) and m. The limits of the HyperCosmos spectrum follow.

Figure 2.1. A segement of the one dimension space embodied in a string of infinite length extending from -∞ to ∞. The center of force is indicated by the filled circle.

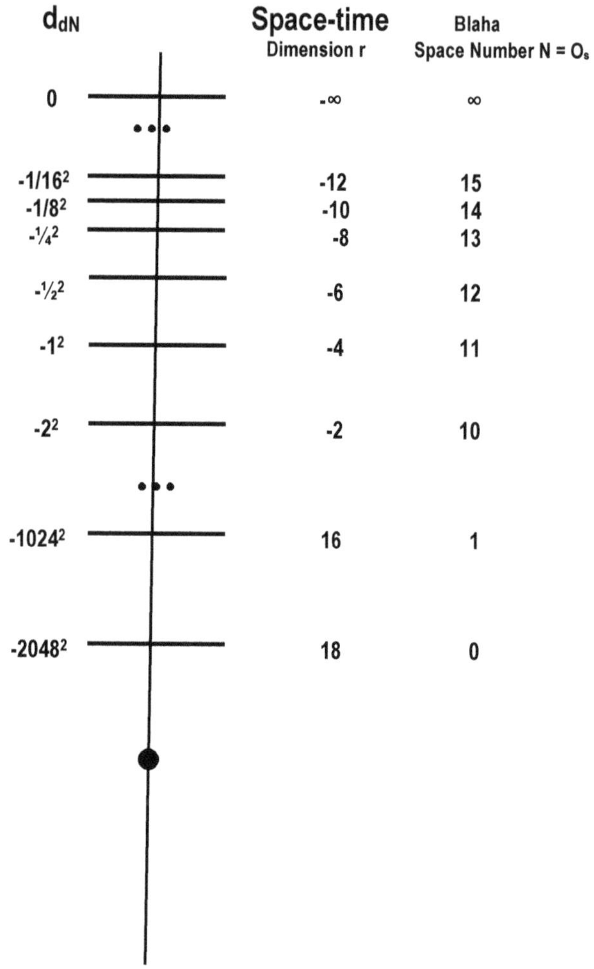

Figure 2.2. The d_{dN} part of the positive energy levels of the UniDimension ProtoCosmos Model (not drawn to scale and a d_{dN} minus sign is not shown). The various quantities match the HyperCosmos spaces spectrum in Fig. 1.1. The energy levels are symbolically depicted. Negative N levels are not shown.

3. Tower ProtoCosmos Model

We now consider the possibility that the ProtoCosmos consists of a zero dimension space-time point. We assume there is a tower of energy values at the point. See Fig. 3.1. These energy values are inputs to a set of Dirac-like equations based on eq. 2.9 for universe wave functions of the form

$$i\gamma^{\mu}\cdot\partial/\partial u^{\mu}|_{r=f(E)}\,\psi(u) = H\psi(u) = E\psi(u) \tag{3.1}$$

for a zero dimension ProtoCosmos space-time viewed as the limit of a one dimension space, and for an r = f(E) universe space-time u.

The mass terms are absorbed into the energy E. Again we make the u space-time dimension dependent on E. The set of energies forms a discrete tower of values. This model determines the HyperCosmos spectrum without dynamics. It constitutes a static model.

We now proceed by analogy with Regge pole trajectories where amplitudes are functions of angular momentum. We choose the set to be numbered by N and we choose the energy values to be[19]

$$E_N = d_{dN} = 2^{22-2N} \tag{3.2}$$

based on the HyperCosmos spaces spectrum dimension arrays. Thus the space-time dimension r is dependent on the energy with[20]

$$\begin{aligned} f(E) = r &= \ln_2(E/16) \\ &= 18 - 2N \end{aligned} \tag{3.3}$$

resulting in

$$\gamma^{\mu}\cdot\partial/\partial u^{\mu}|_{f(E)} = \sum_{\mu=1}^{r} \gamma^{\mu}\cdot\partial/\partial u^{\mu} \tag{3.4}$$

Note the dependence on space-time dimension r

$$r = \ln_2 E - 4 = \ln_2 E/16 \tag{3.5}$$

assuming the eq. 2.17 equations also hold for the Tower ProtoCosmos.

---

[19] Later we will consider the possibility that the energies may be continuous in analogy with Regge pole-like trajectory extensions to continuous angular momentum values.

[20] The factor of 16 figures appears prominently in Fundamental Reference Frame discussions as setting the size of the basic unit of dimensions. See chapters 8, 9 and 10, and also section 10.3, of Blaha (2022c) for details.

3.1 Energy Values determined by r

Eq. 3.3 determines the forms of E and dimension r. If we assume r is an integer then

$$2^r = E/16$$

or

$$E = 2^{r+4}$$

consistent with eq. 2.17. If we express E in terms of N, which is used to number tower elements, then

$$E_N = 2^{r+4} = 2^{22-2N} = d_{dN} \tag{3.6}$$

3.2 Spin – Energy Relation

The space-time dimension of a space is directly related to the spin of a Dirac fermion in the space. The number of rows in a Dirac gamma matrix γ_C satisfies

$$\gamma_C = 2^{r/2} \tag{3.7}$$

In general it is related to fermion spin by

$$\gamma_C = 4s + 2 \tag{3.8}$$

Combining eq. 3.7 and 3.8 with eq 3.6 yields

$$r = \ln_2 (4s + 2)^2$$

$$E_N = 16(4s_N + 2)^2 \tag{3.9}$$

for HyperCosmos space N.

As a function of N we also find the spin relations

$$s = 2^{7-N} - ½ \tag{3.9a}$$

and

$$s_N = 2s_{N+1} + ½ \tag{3.9b}$$

for $N \geq 0$. The values of s are

N	s
9	-¼
8	0
7	½
6	3/2
5	7/2
4	15/2
3	31/2
2	63/2
1	127/2
0	255/2

Table 3.1. Spins of fermion wave functions in HyperCosmos spaces. Higher spin wave functions are acceptable with divergence-free perturbation theories if Quantum Field Theory is formulated using the Two-Tier formalism.

4.2.1 Spin – Amplitude Relation

We thus have an energy – spin relationship similar to the amplitude – angular momentum relation of Regge pole trajectories. Both exhibit a linear relation. The spin and the energy may be treated as continuous values. Then the dimension r would be non-integral in general. The result is a crude analog to dimensional regularization.

Thus one may envision a zero space-time dimension Tower ProtoCosmos where the Tower ProtoCosmos point has a series of allowed values for E. The set of generated spaces have the form of the HyperCosmos spectrum.

Pursuing the Regge pole trajectory analogy we define a ProtoCosmos amplitude trajectory α(s) with

$$\alpha(s) = \sqrt{E} = 4(4s + 2) \tag{3.10}$$

α(s) has a linear dependence on spin s similar to a Regge pole trajectory. At certain spins we find corresponding HyperCosmos spaces. We specify them as functions of Blaha number N:

$$\alpha(s_N) = \sqrt{E_N} = 4(4s_N + 2) \tag{3.11}$$

in analogy with Regge pole trajectory resonances. .

Since the dimension array column lengths d_{cdN} in Fig. 1.1 satisfy

$$d_{cdN} = \sqrt{d_{dN}} = \sqrt{E'} = \alpha(s_N) \tag{3.12}$$

we find

$$d_{cdN} = 16s_N + 8 \tag{3.13}$$

where s_N is the Blaha space N wave function spin.

The Tower ProtoCosmos then has a spaces' trajectory similar to Regge pole trajectories. If we allow N to be continuous then we obtain a continuous spectrum of space-time dimensions r and dimension arrays interspersing non-integer-based spaces within the HyperCosmos spaces of Fig. 1.1. See Figs. 3.1 and 3.2 below.

These considerations raise the possibility of a deeper synthesis of the ProtoCosmos. See chapter 4.

3.3 Universe Wave Function

The universe wave function for the Dirac equation eq. 3.1 is

$$\psi_{2iMN\alpha}(\mathbf{u}, t_u) = \Sigma_s \int d^{r-1}q\, \mathfrak{N}(q)[u(q, s)b_{iMN\alpha}(q,s)e^{-iq\cdot u} + v(q, s)d_{iMN\alpha}(q,s)^{\dagger}e^{iq\cdot u}] \quad (3.14)$$

for i = 1, 2 where $\mathfrak{N}$ is a normalization constant, where M is the mass-energy in the universe space of Blaha number N, r is the universe space-time dimension, and α ranges from 1 through d_{dN}. We assume $q^2 = M^2$ in the universe space. Note $u = (u^0, \mathbf{u})$, and $t_u = u^0$.

The solution eq. 3.14 gives a space fermion of the type described in Blaha (2022c). The range of the symmetry index α from 1 through d_{dN} gives a dimension array of d_{dN} elements that can be subdivided ito a d_{cd} by d_{cd} square array of the type seen in the HyperCosmos spaces spectrum. Universe states can be created from sums over α of the wave functions $\psi_{22MN\alpha}(\mathbf{u}, t_u)$.

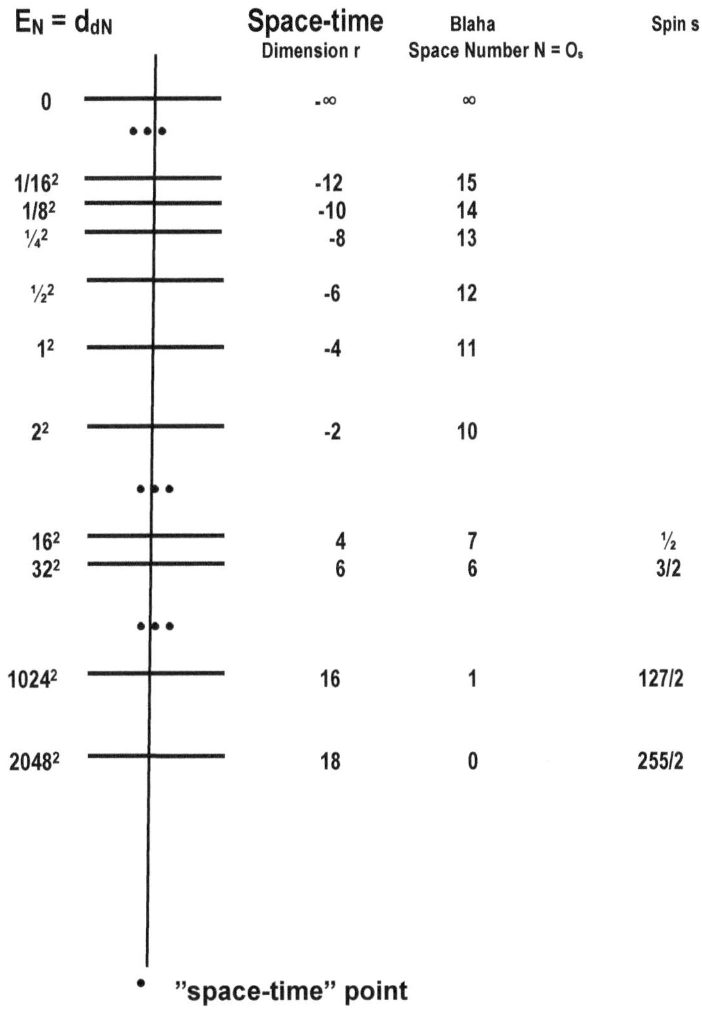

Figure 3.1. The energy levels of a Tower ProtoCosmos. It is not drawn to scale. The various quantities are from the HyperCosmos spaces spectrum in Fig. 1.1. The energy levels are symbolically depicted on the tower. The · represents the Tower ProtoCosmos point space.

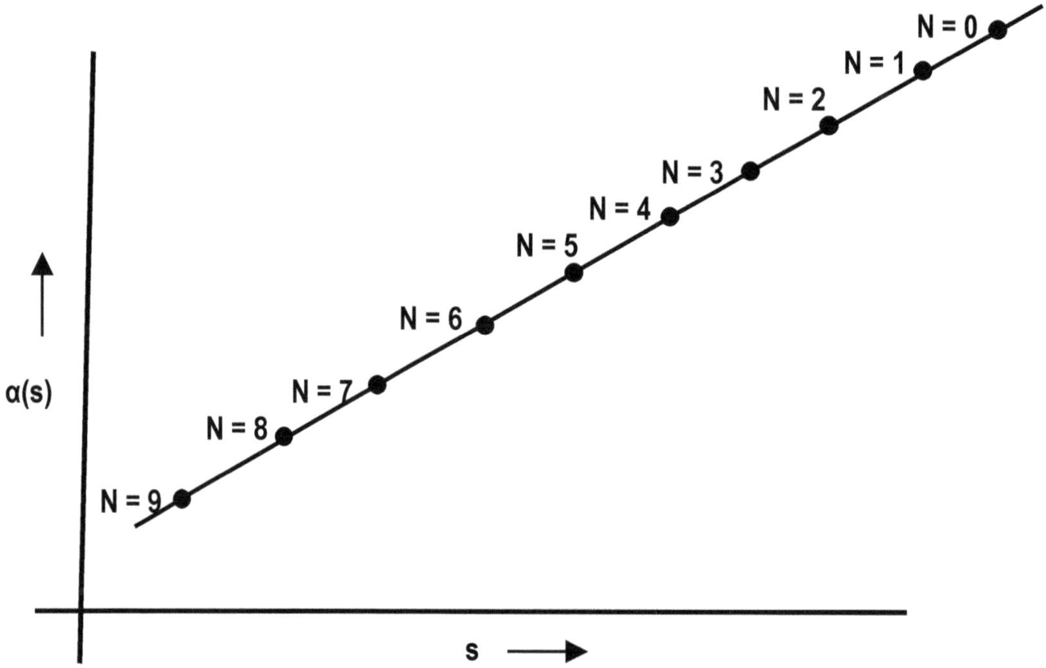

Figure 3.2. Amplitude vs. spin qualitative diagram for eq. 3.10 where $\alpha(s) = \sqrt{E} = 16s + 8$ where the spin is treated as a continuous parameter. Each "dot" corresponds to a HyperCosmos space. The spaces are Blaha numbered as in Fig. 1.1. They correspond to spins for $N \leq 9$ appearing above in Table 3.1 based on eqs. 3.9a and 3.9b.

4. The HyperCosmos Trajectory Spaces as Regge Trajectory Resonances

The Regge trajectory view developed in chapter 3 applies equally well to the spectrum of the UniDimension ProtoCosmos model of chapter 2. In this chapter we seek to deepen the Regge trajectory interpretation of the energy spectrum and its relation to spin.

4.1 Spin Angular Momentum Dependence of the ProtoCosmos Trajectory

First it is important to note that the HyperCosmos trajectory is based on spin[21] rather than angular momentum. Spin, after all, is a form of angular momentum. In the UniDimension ProtoCosmos orbital angular momentum is absent. Thus spin is the sole form of angular momentum.

Based on the Regge pole trajectory analogy we defined a ProtoCosmos amplitude trajectory $\alpha(s)$ with

$$\alpha(s) = \sqrt{E} = 4(4s + 2) \qquad (3.10)$$

$\alpha(s)$ has a linear dependence on spin s similar to a Regge pole trajectory. At certain spins we found corresponding HyperCosmos spaces. We specified them as functions of Blaha number N:

$$\alpha(s_N) = \sqrt{E_N} = 4(4s_N + 2) \qquad (3.11)$$

4.2 Intermediate Spaces of the ProtoCosmos Trajectory

As in Regge theory we may consider points in the trajectory intermediate between the ten spaces in Fig. 3.2. Solving eq. 3.10 above we find

$$s = (\sqrt{E} - 8)/16 \qquad (4.1)$$

If we vary E then the spin s changes accordingly. The number of space-time dimensions r also changes since

$$r = \ln_2 E - 4 = \ln_2 E/16 \qquad (3.5)$$

and

$$r = 4s + 2$$

Thus there is a spin dependence of continuous space-time dimension. (Dimensional regularization in gauge Field Theory also uses non-integral dimensions.)

[21] The lowest fermion spin in a space.

The universes that could be created with intermediate non-integral spin and dimension would have peculiar features that bear further investigation.

4.3 Decay of "Resonances" of the ProtoCosmos Trajectory

Regge trajectory resonances may decay to lower resonances. Similarly it is possible for UniDimension universes to decay to lower dimension universes beneath them. This type of universe decay is based on the quadrupling feature of HyperCosmos spaces' dimension arrays. The Blaha number N space dimension array is composed of the dimension arrays of four spaces of type N + 1. Thus a universe may "decay" along the UniDimension trajectory in Fig. 3.2.

A sexic fermion lagrangian interaction term for universe decay is

$$\mathscr{L} = \overline{\psi}_{2iMN\alpha}(u)\psi_{2iMN\alpha}(u)\,\overline{\psi}_{2iM'N+1\beta}(u)\,\psi_{2iM'N+1\beta}(u)\overline{\psi}_{2iM'N+1\gamma}(u)\psi_{2iM'N+1\gamma}(u)$$

summed over α, β, and γ where M' is the "mass" of the produced universes. Other terms of similar form might be added. This sexic interaction does not introduce any divergences if Two-Tier coordinates are used.

In the UniDimension model there is no time coordinate. Consequently the universe decay transition enables a universe to simultaneously have amplitude for the "undecayed" universe and amplitude for the decayed four universe state. In the two dimension ProtoCosmos model a universe may decay in time to four lower universes.

5. UniDimension Universe Wave Functions

The universe wave function derived in chapter 2

$$\psi_{2iMN\alpha}(\mathbf{u}, t_u) = \Sigma_s \int d^{r-1}q\ \mathfrak{N}(q)[u(q, s)b_{iMN\alpha}(q,s)e^{-iq\cdot u} + v(q, s)d_{iMN\alpha}(q,s)^{\dagger}e^{iq\cdot u}] \quad (2.22a)$$

is a wave function within the universe as described by the chapter 2 passage:

The Spaces of the UniDimension ProtoCosmos

This ProtoCosmos Model has two relevant types of space-times. First there is the one dimension ProtoCosmos "space-time." There are also the space-times of each of the model's eigensolutions. For each energy E_N there is the r space-time dimension wave function of eq. 2.22. There is also an internal r space-time dimension set of coordinates within the dimension array d_{dN}. The number of these internal space-time coordinates is the value $r = 18 - 2N$. Also the f(E) condition sets r for the wave function.

The internal number of space-time dimensions always equals the number of wave function space-time dimensions by construction due to the f(E) condition. Therefore we are allowed to identify the wave function space-time with the internal space-time of each eigensolution. For each eigensolution there is the one dimension ProtoCosmos space and one r dimension internal space-time. The f(E) condition causes the wave function and internal space-times of the model to be the same space-time.

Each universe of each HyperCosmos space has a wave function of the form of eq. 2.22a that exists *within* the universe. Consequently it is possible to second quantize a universe consistently within it. This may be viewed as remedying previous attempts to quantize a universe from "outside" the universe. This quantization is based on a specific metric and does not address the issue of quantizing the metric.

5.1 Role of Universe Wave Function

The universe wave function satisfies

$$i\gamma^{\mu}\cdot\partial/\partial u^{\mu}|_{r=f(E)}\, \psi(u) = H\psi(u) = E\psi(u) \quad (3.1)$$

It has the form of eq. 2.22a above.

The role of this wave function is to specify the interior of the universe. One can specify the density of α type mass-energy in a universe state |U> with

$$\rho_{MN\alpha}(u) = |\langle 0|\psi_{2iMN\alpha}(\mathbf{u}, t_u)|U\rangle|^2 \quad (5.1)$$

5.2 Harmonic Oscillator Universe Binding Potential

Eq. 3.1 above specifies a formless universe with a "free" interior. We can give form to the universe by introducing an interaction potential with binding in the space-like region such as

$$V(u) = - g^2u^2\, \theta(-u^2) \quad (5.2)$$

where $u = (u^{0\,2} - |\mathbf{u}|^2)^{1/2}$. The binding is restricted to the space-like region since the time-like region effectively has "built-in" binding due to the first term in eq. 5.3 below.

Eq. 3.1 above is now modified to

$$[i\gamma^\mu \cdot \partial/\partial u^\mu - g^2 u^2\, \theta(-u^2)]\,|_{r=f(E)}\, \psi(u) = H\psi(u) = E\psi(u) \qquad (5.3)$$

Note the form of u depends on the space-time dimension r.

The Klein-Gordan equivalent to eq. 5.3 has been solved.[22] The solution of eq. 5.3 presumably has a form with a large mass-energy density near u = 0 in the space-like region. Qualitatively one expects

$$\rho(u) \sim A(u)\exp[-B(\mathbf{u}^2 - u^{0\,2})] \qquad (5.4)$$

with a larger density near $|\mathbf{u}| = 0$.

The density distribution of the universe may be modified to simulate a Big Bang followed by expansion by introducing a time dependent term in eq. 5.3.

$$[i\gamma^\mu \cdot \partial/\partial u^\mu + V(u, t)]\,|_{r=f(E)}\, \psi(u) = H\psi(u) = E\psi(u) \qquad (5.3)$$

The Big Bang and universe Hubble expansion are described in Blaha (2021d), (2004) and (2019e).

[22] Stephen Blaha, Phys. Rev. **D12**, 3921 (1975).

6. Derivation of the HyperCosmos Spaces from Creation/Annihilation Operators

The HyperCosmos spectrum of spaces was generated in chapter 2 – UniDimension ProtoCosmos and in chapter 3 – Tower ProtoCosmos. We found the space-time dimensions and the dimension arrays of the ten spaces supporting universe instances. We now perform an alternate derivation that leads to the same spectrum. It is based on the creation and annihilation operators of fermion wave functions and their transformations under General Relativistic transformations in various space-time dimensions.

6.1 GiFT for Particle Creation/Annihilation Operator Transformations

In Blaha (2021i) and (2021j) we developed the features of Generalized Field Theory (GiFT).[23] In this chapter we describe it and apply its formalism to the cases of bosons and fermions. *We will show the HyperCosmos spectrum of spaces naturally emerges.*

We begin by considering two 4-dimension coordinate systems denoted α and β and develop a relation between the quantizations in these coordinate systems for scalar bosons, and for fermions where the spins of the fermions in the respective coordinate systems are the same. Then we consider the fermion case where the energies and spins of states in one coordinate system correspond to superpositions of energy and spin states in other coordinate system.

Since higher dimension space-times appear in the HyperCosmos we develop the relation between quantizations of coordinate systems in higher dimensions.

In the discussions of creation/annihilation operators we will determine features for a given momentum, usually denoted k. The results will typically lead to CASe groups[24] that are infinite tensor products over all values of k in each HyperCosmos space.

Note on PseudoQuantum Theory

Some advantages of PseudoQuantum Field Theory are:

1. For Quantization in any coordinate system in flat or curved space-times with an invariant definition of asymptotic particle states. *This is especially important for the higher dimension spaces of the HyperCosmos.* An n particle asymptotic state in one coordinate system has a unitarily equivalent n particle asymptotic state in any other coordinate system. Therefore particle number is invariant under change of coordinate system. This is important for the HyperCosmos and the Unified SuperStandard Theory in curved space-times. *It is also important for*

[23] GiFT is the Quantum Field Theory based on PseudoQuantum Theory and Two-Tier Theory—both created by this author. PseudoQuantum Field Theory is needed in HyperCosmos Physics. It is needed for proper quantization in arbitrary coordinate systems in higher dimension HyperCosmos spaces. See Blaha (2021i) for details.

[24] See Blaha (2021i) and (2021j) for CASe details.

quantization in higher dimensional spaces such as those of the HyperCosmos. The method was developed in the late 1970's by the author to provide a quantization procedure which supports a unique particle interpretation of states in arbitrary non-static space-times where a different, or a non-global time-like coordinate (Killing vector) exists. PseudoQuantum Field Theory, which we developed in a series of papers and books,[25] also can be formulated in the HyperCosmos Spectrum of spaces. For example, we can use it to implement the Higgs Mechanism to generate particle masses and symmetry breaking.

2. PseudoQuantum Field Theory enables one to define Higgs particle dynamics in such a way that a non-zero vacuum expectation value cleanly separates from the quantum field part of the Higgs fields. This technique can be used in symmetry breaking mechanisms, mass generation, and possible generation of coupling constants as vacuum expectation values.

3. It supports the canonical definition of higher derivative field theories through the use of the Ostrogradski bootstrap. We have used it to construct fourth order theories of the Strong interaction that has color confinement and a linear r potential. We also construct a fourth order theory of Gravity with results analogous to MOND. Chapters 13 and 14.

Scalar Boson GiFT and CASe Creation/Annihilation Transformations

6.1.1 GiFT for Scalar Creation/Annihilation Operator Transformations

We illustrate the procedure by considering the case of a scalar particle in four dimensions[26] and expand on the development given in section 4 of S. Blaha, "The Local Definition of Asymptotic Particle States", IL Nuovo Cimento **49A**, 35 (1979), which we call I. *The PseudoQuantum Field Theory in I, and other1970s papers are, essential for the development of satisfactory GiFT and CASe formulations.* See Blaha (2021i) and (2021j) for additional details.

We associate a scalar particle with two scalar fields: $\varphi_1(x)$ and $\varphi_2(x)$. We choose $\varphi_1(x)$ to have a zero equal time commutator with $d\varphi_1(x)/dx^0$ and $\varphi_2(x)$ to have a conventional equal time commutator with $d\varphi_2(x)/dx^0$. Conceptually $\varphi_1(x)$ is a "classical" field and $\varphi_2(x)$ is a quantum field. A Lagrangian that implements these choices of commutation relations is:

$$\mathcal{L} = \partial^\mu \varphi_1(x)\partial_\mu\varphi_2(x) - \tfrac{1}{2}\,\partial^\mu \varphi_1(x)\partial_\mu\varphi_1(x) - m^2\,\varphi_1(x)\varphi_2(x) + \tfrac{1}{2}\,m^2\,\varphi_1(x)^2 \tag{6.1}$$

$$(\square + m^2)\varphi_1(x) = 0 \tag{6.2}$$

$$(\square + m^2)\varphi_2(x) - (\square + m^2)\varphi_1(x) = 0 \tag{6.3}$$

The canonical momenta are

$$\pi_1 = d\varphi_2(x)/dt - d\varphi_1(x)/dt \tag{6.4}$$

$$\pi_2 = d\varphi_1(x)/dt \tag{6.5}$$

---

[25] See Blaha (2017b) for the discussion of the PseudoQuantum field theory formalism for Higgs particles in our Extended Standard Model. See chapter 20 of Blaha (2017b), and earlier books, for a more detailed view than that presented here.

[26] The case of a space with a higher dimension spacetime is completely analogous. See Appendix 6-A for a description of Blaha's PseudoQuantum Theory and Two-Tier Quantum Theory.

and their equal time commutation relations are

$$[\varphi_i(x), \pi_j(y)] = i\delta_{ij}\delta(\mathbf{x} - \mathbf{y}) \quad (6.6)$$
$$[\varphi_i(x), \varphi_j(y)] = [\pi_i(x), \pi_j(y)] = 0 \quad (6.7)$$

for i, j = 1, 2 implying

$$[\varphi_1(x), d\varphi_1(y)/dt] = 0 \quad (6.8)$$
$$[\varphi_2(x), d\varphi_2(y)/dt] = i\delta^3(\mathbf{x} - \mathbf{y}) \quad (6.9)$$
$$[\varphi_1(x), d\varphi_2(y)/dt] = i\delta^3(\mathbf{x} - \mathbf{y}) \quad (6.10)$$

The *most* general mode expansion of the fields in the coordinate system A with Fourier momentum α is

$$\varphi_{1A}(x) = \Sigma_\alpha\, [a_{1\alpha}\, f_\alpha(x) + a^\dagger{}_{1\alpha}\, f_\alpha{}^*(x)] \quad (6.11)$$
$$\varphi_{2A}(x) = \Sigma_\alpha\, [a_{2\alpha} f_\alpha(x) + a^\dagger{}_{2\alpha}\, f_\alpha{}^*(x)] \quad (6.12)$$

and in coordinate system B with Fourier momentum β is

$$\varphi_{1B}(x) = \Sigma_\beta\, [a_{1\beta}\, f_\beta(x) + a^\dagger{}_{1\beta}\, f_\beta{}^*(x)] \quad (6.13)$$
$$\varphi_{2B}(x) = \Sigma_\beta\, [a_{2\beta}\, f_\beta(x) + a^\dagger{}_{2\beta}\, f_\beta{}^*(x)] \quad (6.14)$$

The creation/annihilation operators of A have the commutation relations:

$$[a_{i\alpha'}, a_\alpha] = 0 \quad (6.15)$$
$$[a^\dagger{}_{i\alpha'}, a^\dagger{}_{j\alpha}] = 0$$
$$[a_{i\alpha}, a^\dagger{}_{j\alpha'}] = (1 - \delta_{ij})\delta^3(\boldsymbol{\alpha} - \alpha') \quad (6.16)$$

for i, j = 1, 2. Two related vacuums are defined: $|0>_1$ and $|0>_2$ by

$$a_{1\alpha}|0>_{2\alpha} = a^\dagger{}_{1\alpha}|0>_{2\alpha} = 0 \quad (6.17)$$
$$a_{2\alpha}|0>_{1\alpha} = a^\dagger{}_{2\alpha}|0>_{1\alpha} = 0$$
$$a_{i\alpha}|0>_{i\alpha} \neq 0 \qquad a^\dagger{}_{i\alpha}|0>_{i\alpha} \neq 0 \quad (6.18)$$

for i = 1, 2 and all α and α′. And similarly for the creation/annihilation operators of B with α replaced with β.

The general coordinate system B Fourier expansion in terms of A creation/annihilation operators is

$$\varphi_{1B}(x) = \Sigma_\beta\, \Sigma_\alpha\, [(c_{11}a_{1\alpha} + c_{12}a_{2\alpha} + C_{11}a^\dagger{}_{1\alpha} + C_{12}a^\dagger{}_{2\alpha})f_\beta(x) + \\ + (c'_{11}a^\dagger{}_{1\alpha} + c'_{12}a^\dagger{}_{2\alpha} + C'_{11}a_{1\alpha} + C'_{12}a_{2\alpha})f_\beta{}^*(x)] \quad (6.19)$$

$$\varphi_{2B}(x) = \Sigma_\beta\, \Sigma_\alpha\, [(c_{21}a_{1\alpha} + c_{22}a_{2\alpha} + C_{21}a^\dagger{}_{1\alpha} + C_{22}a^\dagger{}_{2\alpha})\, f_\beta(x) + \\ + (c'_{21}a^\dagger{}_{1\alpha} + c'_{22}a^\dagger{}_{2\alpha} + C'_{21}a_{1\alpha} + C'_{22}a_{2\alpha})f_\beta{}^*(x)] \quad (6.20)$$

where the c_{ij} and C_{ij}, and c'_{ij} and C'_{ij} are all functions of α.

We define sets of Bogoliubov transformations B_1 and B_2 that give

$$a_{i\beta} = \Sigma_\alpha (f_\beta, f_\alpha) B_{2\alpha} B_{1\alpha} a_{i\alpha} (B_{1\alpha} B_{2\alpha})^{-1} = \Sigma_\alpha (f_\beta, f_\alpha)[c_{i1} a_{1\alpha} + c_{i2} a_{2\alpha} + C_{i1} a^\dagger_{1\alpha} + C_{i2} a^\dagger_{2\alpha}] \quad (6.21)$$

$$a^\dagger_{i\beta} = \Sigma_\alpha (f^*_\beta, f^*_\alpha) B_{2\alpha} B_{1\alpha} a^\dagger_{i\alpha} (B_{1\alpha} B_{2\alpha})^{-1} = \Sigma_\alpha (f^*_\beta, f^*_\alpha)[c'_{i1} a^\dagger_{1\alpha} + c'_{i2} a^\dagger_{2\alpha} + C'_{i1} a_{1\alpha} + C'_{i2} a_{2\alpha}] \quad (6.22)$$

6.1.1.1 The B_1 Transformation

The B_1 transformation has the form:

$$B_{1\alpha}(x_1, x_2) = \exp[x_1 \Gamma^1_{3\alpha}] \exp[x_2 \Gamma^1_{2\,\alpha}] \quad (6.23)$$

where the Hermitian operators $\Gamma^1_{1i\alpha}$ are

$$\Gamma^1_{3\alpha} = (a^\dagger_{2\alpha} a_{1\alpha} + a_{2\alpha} a^\dagger_{1\alpha})/2 \quad (6.24)$$
$$\Gamma^1_{2\alpha} = i(a^\dagger_{2\alpha} a^\dagger_{1\alpha} - a_{2\alpha} a_{1\alpha})/2 \quad (6.25)$$

We also define

$$\Gamma^1_{1\alpha} = -(a^\dagger_{2\alpha} a^\dagger_{1\alpha} + a_{2\alpha} a_{1\alpha})/2 \quad (6.26)$$

and

$$\Gamma^1_{4\alpha} = i(a^\dagger_{2\alpha} a_{1\alpha} - a_{2\alpha} a^\dagger_{1\alpha})/2 \quad (6.27)$$

Together the three operators satisfy su(1,1) algebra commutation relations:

$$[\Gamma^1_{1\alpha}, \Gamma^1_{2\alpha'}] = -i\delta_{\alpha\alpha'} \Gamma^1_{3\alpha} \qquad [\Gamma^1_{2\alpha}, \Gamma^1_{3\alpha'}] = i\delta_{\alpha\alpha'} \Gamma^1_{1\alpha} \qquad [\Gamma^1_{3\alpha}, \Gamma^1_{1\alpha'}] = i\delta_{\alpha\alpha'} \Gamma^1_{2\alpha} \quad (6.28)$$

They transform $a_{i\alpha}$ and $a^\dagger_{i\alpha}$ to terms of the form $c_{i\alpha} a_{i\alpha} + c_{i\alpha}' a^\dagger_{i\alpha}$ where $c_{i\alpha}$ and $c_{i\alpha}'$ are constants.

Some properties of the su(1,1) algebra are detailed in S. Blaha, "The Local Definition of Asymptotic Particle States", IL Nuovo Cimento **49A**, 35 (1979).

6.1.1.2 The B_2 Transformation

The B_2 transformation has the form:

$$B_{2\alpha}(x_1, x_2) = \exp[x_1 \Gamma^2_{3\alpha}] \exp[x_2 \Gamma^2_{2\,\alpha}] \quad (6.29)$$

where the Hermitian operators $\Gamma^2_{1i\alpha}$ are

$$\Gamma^2_{3\alpha} = (a^\dagger_{1\alpha} a_{1\alpha} + a_{1\alpha} a^\dagger_{1\alpha})/2 \quad (6.30)$$
$$\Gamma^2_{2\alpha} = (a^\dagger_{2\alpha} a_{2\alpha} + a_{2\alpha} a^\dagger_{2\alpha})/2 \quad (6.30)$$
$$\Gamma^2_{1\alpha} = (a^\dagger_{2\alpha} a^\dagger_{2\alpha} + a_{2\alpha} a_{2\alpha})/2 \quad (6.31)$$

We also define the Hermitian operators

$$\Gamma^2_{4\alpha} = (a^\dagger_{1\alpha} a^\dagger_{1\alpha} + a_{1\alpha} a_{1\alpha})/2 \quad (6.32)$$

and

$$\Gamma^2_{5\alpha} = i(a^\dagger_{2\alpha}a^\dagger_{2\alpha} - a_{2\alpha}a_{2\alpha})/2 \quad (6.33)$$
$$\Gamma^2_{6\alpha} = i(a^\dagger_{1\alpha}a^\dagger_{1\alpha} - a_{1\alpha}a_{1\alpha})/2 \quad (6.34)$$

They transform $a_{1\alpha}$ and $a^\dagger_{1\alpha}$ to have additional terms of the form $c_\alpha a_{2\alpha} + c_\alpha' a^\dagger_{2\alpha}$ where c_α and c_α' are constants; and they transform $a_{2\alpha}$ and $a^\dagger_{2\alpha}$ to have additional terms of the form $c_\alpha a_{1\alpha} + c_\alpha' a^\dagger_{1\alpha}$.

The complete set of Γ operators are part of an su(1,1) group. This group also appears in Twistor Theory. A relation/unification of the group of these Bogoliubov transformations and Twistor Theory may eventually emerge.

The transformation $B_{2\alpha}B_{1\alpha}$ generates the maps of eqs. 6.21 and 6.22. As we will see next the key terms for the transition between annihilation/creation operators of two coordinate systems, A and B, have the form:

$$[c'_{i1}a^\dagger_{1\alpha} + c'_{i2}a^\dagger_{2\alpha} + C'_{i1}a_{1\alpha} + C'_{i2}a_{2\alpha}] \quad (6.35)$$

if we construct states based on $a^\dagger_{2\beta}$ and $a^\dagger_{2\alpha}$ with the vacuum $|0>_2$. Note there are four constants, which are functions of α, in eq. 6.35. These constants are determined by the four parameters in $B_{2\alpha}B_{1\alpha}$. Thus the transformation between coordinate systems embodied in $B_{2\alpha}B_{1\alpha}$ determines the constants in eq. 6.35.

6.1.2 States in the Respective Coordinate Systems

The B states are superpositions of the A states as eq. 6.21 and 6.22 demonstrate. The one particle state in the B coordinate system is

$$a^\dagger_{2\beta}\,|0>_2 = \Sigma_\alpha\,(f^*_\beta, f^*_\alpha)[c'_{21}a^\dagger_{1\alpha} + c'_{22}a^\dagger_{2\alpha} + C'_{21}a_{1\alpha} + C'_{22}a_{2\alpha}]\,|0>_2$$

$$= \Sigma_\alpha\,(f^*_\beta, f^*_\alpha)[c'_{22}a^\dagger_{2\alpha} + C'_{22}a_{2\alpha}]\,|0>_2 \quad (6.36)$$

where the $a^\dagger_{2\alpha}$ term generates a one particle A state, and the $a_{2\alpha}$ term generates an A particle negative energy state. The two particle B coordinate system state

$$a^\dagger_{2\beta}\, a^\dagger_{2\beta}\,|0>_2$$

generates a superposition in momenta α) of two A particles, two A negative energy particles and an A vacuum state term.

6.2 Scalar Particle Creation/Annihilation Space (CASe)

The set of creation and annihilation operators a_1, a_2, $a^\dagger_1$, $a^\dagger_2$ form an operator basis for spaces, for each momentum value α.[27] In the scalar particle case in four space-time dimensions we define a four dimension space that we call the su(1,1)

[27] **Note this calculation requires the use of PseudoQuantum fields.** Two fields × two operators (an a and $a^\dagger$ per field) = 4 real operators. Thus 4 real-valued coordinates. . The fundamental su(1,1) representation has 4 real-valued coordinates. Match.

Creation/Annihilation Space or the CASe(1,1) space. It supports an su(1,1) representation for each α. We call su(1,1) a GiFT transformation group. Defining basis "ortho" vectors $a_\alpha = (a^\dagger_{1\alpha}, a^\dagger_{2\alpha}, a_{1\alpha}, a_{2\alpha})$ we specify complex coordinates for an irreducible su(1,1):

$$x_\alpha = (x_{1\alpha}{}^0, x_{2\alpha}{}^0, x_\alpha{}^1, x_{2\alpha}{}^1) \tag{6.37}$$

Defining the metric

$$ds^2 = |dx_{1\alpha}{}^0|^2 + |dx_{2\alpha}{}^0|^2 - |dx_{1\alpha}{}^1|^2 - |dx_{2\alpha}{}^1|^2 \tag{6.38}$$

we note that there is an su(1,1) group invariance.

Thus for scalar particles an su(1,1) symmetry exists for each Fourier component α in a space-time of any number of space-time dimensions.

The vectors in the space have the form

$$a_\alpha \cdot x_\alpha = x_{1\alpha}{}^0 a^\dagger_{1\alpha} + x_{2\alpha}{}^0 a^\dagger_{2\alpha} + x_\alpha{}^1 a_{1\alpha} + x_{2\alpha}{}^1 a_{2\alpha} \tag{6.39}$$

The "ortho" vectors a_α form an algebra reminiscent of hypercomplex quaternions. The lack of complete commutativity is not a source for disquiet when one considers the anti-commutativity of quaternion, octonion and sedenion multiplication. It can be viewed positively as a step in the direction of a larger unified theory such as

$$\text{Spinor Space} + \text{CASe(n,m)} \tag{6.40}$$

seen later. CASe is enlarged by spin for fermions in the various space-time dimensions.

See chapter 6 of Blaha (2022c) for a discussion of CASe groups for bosons.

Note: A General Relativistic transformation between coordinate systems with Killing vectors may not induce a GiFT transformation. A General Relativistic transformation between a static and non-static coordinate system does induce a GiFT CASe transformation.

Fermion GiFT and CASe Creation/Annihilation Transformations

6.3 GiFT for *Fermion* Creation/Annihilation Operator Transformations

The GiFT formalism developed in Blaha (2021i) and in section 4 of S. Blaha, "The Local Definition of Asymptotic Particle States", IL Nuovo Cimento **49A**, 35 (1979), which we call I, is directly extendable to fermions. We begin by considering 4-dimension space-time coordinate systems. Following the framework developed in section 4 of I we define fermion fields $\psi_1(x)$ and $\psi_2(x)$ in a coordinate system A labeled with α as

$$\psi_{1A}(x) = \Sigma_{\alpha,s}[b_{1\alpha s} u_{\alpha s} f_\alpha(x) + d^\dagger_{1\alpha s} v_{\alpha s} f_\alpha{}^*(x)] \tag{6.41}$$

$$\psi_{2A}(x) = \Sigma_{\alpha,s}[b_{2\alpha s} u_{\alpha s} f_\alpha(x) + d^\dagger_{2\alpha s} v_{\alpha s} f_\alpha{}^*(x)] \tag{6.42}$$

and in another coordinate system B labeled with β as

$$\psi_{1B}(x) = \Sigma_{\beta,s}[b_{1\beta s}u_{\beta s}\, g_\beta(x) + d^\dagger_{1\beta s}v_{\beta s}\, g_\beta{}^*(x)] \qquad (6.43)$$

$$\psi_{2B}(x) = \Sigma_{\beta,s}[b_{2\beta s}u_{\beta s}\, g_\beta\, (x) + d^\dagger_{2\beta s}v_{\beta s}\, g_\beta{}^*(x)] \qquad (6.44)$$

where $f_\alpha(x)$ and $g_\beta(x)$ are Fourier components.

To begin, we assume the spin sectors transform independently. We also anticipate the Fourier expansions are related in a manner. Then for spin value s we have

$$b_{1\beta s} = \Sigma_{\alpha,x}\, (g_\beta,\, f_\alpha)\; u^\dagger{}_{\beta s}u_{\alpha s}\, (c_{11s}b_{1\alpha s} + c_{12s}b_{2\alpha s} + C_{11s}b^\dagger{}_{1\alpha s} + C_{12s}b^\dagger{}_{2\alpha s}) \qquad (6.45a)$$

$$b_{2\beta s} = \Sigma_{\alpha,x}\, (g_\beta,\, f_\alpha)\; u^\dagger{}_{\beta s}u_{\alpha s}\, (c_{21s}b_{1\alpha s} + c_{22s}\, b_{2\alpha s} + C_{21s}b^\dagger{}_{1\alpha s} + C_{22s}b^\dagger{}_{2\alpha s}) \qquad (6.45b)$$

$$d^\dagger{}_{1\beta s} = \Sigma_{\alpha,x}\, (g^*{}_\beta\, f^*{}_\alpha)\; v^\dagger{}_{\beta s}v_{\alpha s}\, (c'_{11s}d^\dagger{}_{1\alpha s} + c'_{12s}d^\dagger{}_{2\alpha s} + C'_{11s}d^\dagger{}_{1\alpha s} + C'_{12s}d^\dagger{}_{2\alpha s}) \qquad (6.46a)$$

$$d^\dagger{}_{2\beta s} = \Sigma_{\alpha,x}\, (g^*{}_\beta\, f^*{}_\alpha)\; v^\dagger{}_{\beta s}v_{\alpha s}\, (c'_{21s}d^\dagger{}_{1\alpha s} + c'_{22s}d^\dagger{}_{2\alpha s} + C'_{21s}d^\dagger{}_{1\alpha s} + C'_{22s}d^\dagger{}_{2\alpha s}) \qquad (6.46b)$$

generalizing eq. 118 of I where the c_{ij} are all constants and where x is a space-like surface. The inner products have the form

$$\Sigma_x\, (g_\beta,\, f_\alpha) = i\; \Sigma_x\, g^*{}_\beta \overleftrightarrow{\partial}_0 f_\alpha \qquad (6.46c)$$

where the subscript "0" indicates the time-like coordinate. For example, in rectilinear coordinates

$$\Sigma_x\, (g_{k'},\, f_k) = i\; \Sigma_x\, g^*{}_k \overleftrightarrow{\partial}_0 f_k = \delta^3(\mathbf{k' - k}) \qquad (6.46d)$$

for momenta k and k′.

The non-zero anti-commutation relations of the creation/annihilation operators are:

$$\{b_{i\alpha s}(\alpha),\, b^\dagger{}_{j\alpha s'}(\alpha')\} = \{d_{i\alpha s}(\alpha\,),\, d^\dagger{}_{j\alpha s'}(\alpha')\} = (\,1 - \delta_{ij})\delta_{ss'}\; \delta^3(\boldsymbol{\alpha} - \boldsymbol{\alpha}') \qquad (6.47)$$
$$\{b_{i\beta s}(\beta),\, b^\dagger{}_{j\beta s'}(\beta')\} = \{d_{i\beta s}(\beta),\, d^\dagger{}_{j\beta s'}(\,\beta')\} = (\,1 - \delta_{ij})\delta_{ss'}\; \delta^3(\boldsymbol{\beta} - \boldsymbol{\beta}')$$

Note the transformations of eqs. 6.45-46 superimpose "momenta" in relating the coordinate system quantizations. But they do not superimpose spins.[28] As a result the transformations can be factored into two sets: one set for spin up and one set for spin down.

6.3.1 Transformations of the b and b† Operators without Spin Mixing or b – d Mixing

The sets of Bogoliubov transformations B_1 and B_2 for either spin up or down b creation/annihilation operators are defined by

[28] We consider spin mixing later.

$$B_{1\alpha}(x_1, x_2) = ex[x_1\Gamma^1{}_{3\alpha}] \exp[x_2\Gamma^1{}_{2\,\alpha}] \tag{6.48}$$

where the Hermitian operators $\Gamma^1{}_{1i\alpha}$ are

$$\Gamma^1{}_{3\alpha}(b_{1\alpha s}, b_{2\alpha s}) = (b^\dagger{}_{2\alpha s}b_{1\alpha s} + b_{2\alpha s}b^\dagger{}_{1\alpha s})/2 \tag{6.49}$$
$$\Gamma^1{}_{2\alpha}(b_{1\alpha s}, b_{2\alpha s}) = i(b^\dagger{}_{2\alpha s}b^\dagger{}_{1\alpha s} - b_{2\alpha s}b_{1\alpha s})/2 \tag{6.50}$$

We also define

$$\Gamma^1{}_{1\alpha}(b_{1\alpha s}, b_{2\alpha s}) = -(b^\dagger{}_{2\alpha s}b^\dagger{}_{1\alpha s} + b_{2\alpha s}b_{1\alpha s})/2 \tag{6.51}$$

and

$$\Gamma^1{}_{4\alpha}(b_{1\alpha s}, b_{2\alpha s}) = -(b^\dagger{}_{2\alpha s}b_{1\alpha s} + b_{2\alpha s}b^\dagger{}_{1\alpha s})/2 \tag{6.52}$$

The B_2 transformation has the form:

$$B_{2\alpha}(x_1, x_2) = \exp[x_1\Gamma^2{}_{3\alpha}] \exp[x_2\Gamma^2{}_{2\,\alpha}] \tag{6.53}$$

where the Hermitian operators $\Gamma^2{}_{1i\alpha}$ are

$$\Gamma^2{}_{3\alpha} = \Gamma^1{}_{3\alpha}(b_{1\alpha s}, b_{1\alpha s}) \tag{6.54}$$
$$\Gamma^2{}_{2\alpha} = \Gamma^1{}_{3\alpha}(b_{2\alpha s}, b_{2\alpha s}) \tag{6.55}$$
$$\Gamma^2{}_{1\alpha} = \Gamma^1{}_{1\alpha}(b_{2\alpha s}, b_{2\alpha s}) \tag{6.56}$$

We also define the Hermitian operators

$$\Gamma^2{}_{4\alpha} = \Gamma^1{}_{1\alpha}(b_{1\alpha s}, b_{1\alpha s}) \tag{6.57}$$
$$\Gamma^2{}_{5\alpha} = \Gamma^1{}_{2\alpha}(b_{2\alpha s}, b_{2\alpha s}) \tag{6.58}$$
$$\Gamma^2{}_{6\alpha} = \Gamma^1{}_{2\alpha}(b_{1\alpha s}, b_{1\alpha s}) \tag{6.59}$$

The above set of operators generate the su(1,1) transformations to the forms of eqs. 6.46. The su(1,1) representation has four real components for b_i, and $b_i^\dagger$ for i = 1, 2.

6.3.2 Transformations of the d and $d^\dagger$ Operators without Spin Mixing or b – d Mixing

The operators for "holes" have an analogous form to the b operators. These operators are part of the operators of another su(1, 1).

6.4 Operators of Types b and d *with* Spin Mixing and b – d Mixing

These operators, plus those of section 6.1, are part of the operators of an su(4,4), which contains operators mixing spins, and b and d operators. The su(4, 4) fundamental representation, which mixes spins, and b's and d's, has eight real components for b_i, $b_i^\dagger$ d_i, and $d_i^\dagger$ for i = 1, 2 for each spin.

$$su(2, 2)\otimes su(2, 2) \rightarrow su(4, 4) \tag{6.60}$$

We can anticipate that certain General Relativistic transformations may cause spin mixing. Some typical su(4, 4) Creation/Annihilation Space (CASe) transformations are (in the notation of section 4.1)

$$B_{1\alpha}(x_1, x_2) = \exp[x_1\Gamma^1{}_{3\alpha}]\exp[x_2\Gamma^1{}_{2\alpha}] \tag{6.61}$$

with

$$\Gamma^1{}_{3\alpha}(b_{1\alpha s}, b_{2\alpha s'})$$
$$\Gamma^1{}_{2\alpha}(b_{1\alpha s}, b_{2\alpha s'})$$
$$\Gamma^1{}_{1\alpha}(b_{1\alpha s}, b_{2\alpha s'})$$
$$\Gamma^1{}_{4\alpha}(b_{1\alpha s}, b_{2\alpha s'})$$

where $s \neq s'$.
The B_2 transformation has the form:

$$B_{2\alpha}(x_1, x_2) = \exp[x_1\Gamma^2{}_{3\alpha}]\exp[x_2\Gamma^2{}_{2\alpha}] \tag{6.62}$$

where

$$\Gamma^2{}_{3\alpha} = \Gamma^1{}_{3\alpha}(b_{1\alpha s}, b_{1\alpha s'})$$
$$\Gamma^2{}_{2\alpha} = \Gamma^1{}_{3\alpha}(b_{2\alpha s}, b_{2\alpha s'})$$
$$\Gamma^2{}_{1\alpha} = \Gamma^1{}_{1\alpha}(b_{2\alpha s}, b_{2\alpha s'})$$
$$\Gamma^2{}_{4\alpha} = \Gamma^1{}_{1\alpha}(b_{1\alpha s}, b_{1\alpha s'})$$
$$\Gamma^2{}_{5\alpha} = \Gamma^1{}_{2\alpha}(b_{2\alpha s}, b_{2\alpha s'})$$
$$\Gamma^2{}_{6\alpha} = \Gamma^1{}_{2\alpha}(b_{1\alpha s}, b_{1\alpha s'})$$

and $s \neq s'$.

6.6.1 Four Space-Time Dimension Case

In four dimensions there are two spin values, up and down. Consequently the combination of spin mixed transformations gives su(2, 2) separately for the b's and the d's for i =1, 2. The su(4, 4) representation has eight real-valued components for the b_i, and $b_i^\dagger$, and d_i, and $d_i^\dagger$, for i = 1, 2 taking account of both spins.[29] Thus the CASe group for four dimension space-time is

$$\text{su}(4, 4) \tag{6.63}$$

The fundamental representation of su(4,4) group has a 16 × 16 matrix form in a 16 real dimension space. Its vectors are analogous to the set of creation/annihilation operators:

$$(b_{1\uparrow}, b_{2\uparrow}, b^\dagger{}_{1\uparrow}, b^\dagger{}_{2\uparrow}, b_{1\downarrow}, b_{2\downarrow}, b^\dagger{}_{1\downarrow}, b^\dagger{}_{2\downarrow}, d_{1\uparrow}, d_{2\uparrow}, d^\dagger{}_{1\uparrow}, d^\dagger{}_{2\uparrow}, d_{1\downarrow}, d_{2\downarrow}, d^\dagger{}_{1\downarrow}, d^\dagger{}_{2\downarrow}) \tag{6.64}$$

Note that the space has four quartets of vectors. This arrangement will be found to persist to the level of the dimension and fermion arrays of NEWQUeST, the octonion

---

[29] **Note this calculation again requires the use of PseudoQuantum fields.** The four operators (b_i, $b_i^\dagger$, d_i, $d_i^\dagger$) for i = 1,2 and two spin orientations yield 16 real-valued "operator" coordinates. . The fundamental su(4, 4) representation has 16 real-valued coordinates. Thus a match.

If the PseudoQuantum formalism were not used then quantities such as the number of creation/annihilation operators would generally be cut in half. Then the UST would have to be modified by cutting the number of dimensions by a factor of 2 with the result that the number of fundamental fermions would be reduced by a factor of two. The simplest modification of UST in this case would be to remove the Dark fundamental fermions in the UST.

spectrum space of our universe. The subdivisions appear in the known fermion and internal symmetry fundamental representations.

If we consider space-time dimension 4 transformations with spin and b – d mixing then the form of the transformed creation/annihilation (eq. 6.45a) operator becomes

$$b_{1\beta s} = \Sigma_{\alpha,x,s}\,(g_\beta, f_\alpha)\, u^\dagger_{\beta s} u_{\alpha s}\,(c_{11s} b_{1\alpha s} + c_{12s} b_{2\alpha s} + C_{11s} b^\dagger_{1\alpha s} + C_{12} b^\dagger_{2\alpha s} + \\ + c'_{11s} d_{1\alpha s} + c'_{12s} d_{2\alpha s} + C'_{11s} d^\dagger_{1\alpha} + C'_{12s} d^\dagger_{2\alpha})$$

with similar forms for the other transformed operators.

6.4.2 Higher Space-Time Dimension Case

In a space-time of dimension r there are $2^{r/2-1}$ spin values. As a result there is an su($2^{r/2}$, $2^{r/2}$) CASe group containing the direct product subgroup:

$$\bigotimes_{s=1}^{2^{r/2-1}} su(2,2)_s \tag{6.65}$$

where s labels the spin value. Each spin value has an associated su(2, 2) subgroup. For example, for spin ½ there are two subgroups (see above) that combine in an su(4, 4) CASe group.

The combined mixing yields the CASe group for a space-time of r dimensions:

$$su(2^{r/2}, 2^{r/2}) \tag{6.66}$$

with $2^{r/2+2}$ real components for the b_i, $b_i^\dagger$, d_i, and $d_i^\dagger$ for PseudoQuantum fields labeled i = 1, 2 and taking account of all spins.

It is understood that the su($2^{r/2}$, $2^{r/2}$) groups are factors in an infinite tensor product over all momenta.

The number of b's and d's for each momentum, taking account of the PseudoQuantum formalism and spins, is

$$\text{Number of d's and b's} = 2^{r+4}$$

6.4.3 Size of the Fundamental Representations of su($2^{r/2}$, $2^{r/2}$)

The matrices of the fundamental representations of su($2^{r/2}$, $2^{r/2}$) are square with $2^{r/2+2}$ rows and $2^{r/2+2}$ columns of real-valued elements. The number of real-valued elements in a matrix is 2^{r+4}. If we use the relation in Blaha (2022c) between the space-time dimensions r and the Blaha number N:

$$N = Os = ½(18 - r) \tag{2.30}$$

then we find the number of real-valued elements for an su($2^{r/2}$, $2^{r/2}$) fundamental representation matrix equals the size of the corresponding dimension array d_{dN} in Fig. 1.1:

$$d_{dN} = 2^{22-2N} = 2^{r+4} \tag{6.67}$$
$$= \text{Number of b's and d's for each momentum}$$

Eq. 6.67 is important in the derivation of the HyperCosmos spectrum.

6.5 The HyperCosmos Spectrum of Spaces *Derived*

The derivation of the HyperCosmos spectrum based on the above discussion begins by noting that the space-time dimension r must be even numbered or eq. 6.66 and the number of b's and d's would both be fractional, and thus inconsistent with the form of second quantized fermion wave functions seen earlier.

Secondly, the lowest *physical* space-time dimension is r = 0. Thus the physical spectrum of HyperCosmos spaces consists of spaces with an even number of space-time dimensions starting at r = 0. The form of the expression for the number of b's and d's in eq. 6.67 limits N to $N \le 9$ since we assume $r \ge 0$.

The lower bound on N: $N \ge 0$ follows from the existence of a lower bound on positive energies in the Hydrogen-like atom of the ProtoCosmos.

Thus we obtain the HyperCosmos spectrum of spaces in Figs. 1.1 and 6.1 with:

1. A lower bound: $N \ge 0$
2. A series of spaces with space-time dimensions decreasing by two space by space.
3. An upper bound: $N \le 9$.

Note that for N = 7 we find r = 4 space-time dimensions and 256 dimensions in the dimension space d_{dN} as seen in QUeST and the Unified SuperStandard Theory (UST)—our universe!

Thus our creation/annihilation operator derivation yields the HyperCosmos spectrum of Figs. 1.1 and 6.1. We may then view the creation/annihilation operator derivation as determining the choice of parameters.

6.6 The Fermion CASe Groups

The CASe groups of fermions of the universes of The HyperCosmos are determined above in eq. 6.66. See Fig. 6.1.

A General Relativistic transformation to a non-static coordinate system with a different, or without, a Killing vector causes fermion wave function creation/annihilation operators to be CASe transformed in every HyperCosmos space (universe). The correspondence between General Relativistic transformations and the CASe transformations of GiFT reflects the unity of General Relativity and Quantum Theory.

We take the fermion based derivation of the HyperCosmos spectrum from creation/annihilation operator enumeration as a decisive corroboration of the HyperCosmos derivation in chapters 2 and 3.

6.7 The N > 9 Spectrum Spaces of Negative Space-time Dimension

The spectrum in Fig. 6.1 contains spaces with negative space-time dimensions. These spaces have fractional totals for dimensions d_{dN}. We discuss these spaces in chapter 12 of Blaha (2022c) where we show negative space-time dimensions have a reasonable mathematical/physical justification. Fractionated particles and dimensions can also be understood physically.

6.8 Anti-Spaces and Anti-Universes

Section 2.10 of Blaha (2022c) discusses anti-spaces and anti-universes.

CASe Groups of the HyperCosmos Spaces (Universes)

	Blaha Space Number	Cayley-Dickson Number	Cayley Number	Dimension Array column length	Space-time-Dimension	CASe Group $su(2^{r/2},2^{r/2})$
	$N = o_s$	n	d_c	d_{dN}	r	CASe
	0	10	1024	2048^2	18	su(512,512)
	1	9	512	1024^2	16	su(256,256)
	2	8	256	512^2	14	su(128,128)
	3	7	128	256^2	12	su(64,64)
	4	6	64	128^2	10	su(32,32)
	5	5	32	64^2	8	su(16,16)
	6	4	16	32^2	6	su(8,8)
	7	**3**	**8**	$\mathbf{16^2}$	**4**	**su(4,4)**
	8	2	4	8^2	2	su(2,2)
	9	1	2	4^2	0	su(1,1)
EXTENSION:						
	10	0	1	2^2	-2	U(1)
	11	-1	½	1^2	-4	U(½)
	12	-2	¼	$½^2$	-6	•
	13	-3	1/8	$¼^2$	-8	•
	14	-4	1/16	$1/8^2$	-10	•

•
•
•

Figure 6.1. The CASe groups of the HyperCosmos spaces.

Appendix 6-A. Two-Tier and PseudoQuantum Theory Formalisms

The PseudoQuantum field Theory appears in Blaha (2021j) as well as.

S. Blaha, Phys. Rev. **D17**, 994 (1978).
S. Blaha, "The Local Definition of Asymptotic Particle States", IL Nuovo Cimento **49A**, 35 (1979).
S. Blaha, "New Framework for Gauge Field Theories", IL Nuovo Cimento **49A**, 113 (1979).

The PseudoQuantum Theory of Color Confinement appears in

S. Blaha, Phys. Rev. **D10**, 4268 (1974).
S. Blaha, Phys. Rev. D**11**, 2921 (1975).

PseudoQuantum Formalism

We are familiar with Quantum Theory, as it is usually developed, in our universe. In universes with higher space-time dimensions there is a need for a more expansive Quantum Field Theory. We presented this theory in earlier papers and books on Two-Tier Quantum Theory[30] and PseudoQuantum Field Theory.[31] Now it is GiFT. It resolved all divergences in ElectroWeak and Strong interaction theories. It also eliminated divergences in other types of Quantum Field Theories including theories with higher order derivatives and four fermion interactions.

In this book we focus on the higher space-time dimension spaces (universes) of the HyperCosmos. In these universes we find Two-Tier Quantum Theory is *absolutely* necessary to avoid divergences in perturbation theory calculations. The universes of the HyperCosmos have Feynman propagators and Perturbation Theory terms with integrations of the form:

$$\int d^n k \; f(k) \qquad (6\text{-A.1})$$

where n = 4, 8, 10, 12, 14, 16, and 18 (and also a less interesting case n = 2.)

In higher space-time dimensions Two-Tier quantization is essential. For example, in eight space-time dimensions the second order single fermion loop vacuum polarization is sextic divergent: $\int d^8k/k^2 \sim k^6$. Two-Tier Quantum Theory eliminates

[30] See Blaha (2005a) *Quantum Theory of the Third Kind.*
[31] Blaha (2018e) and earlier books.

divergences in Perturbation Theory with exponentiated Gaussian quadratic terms in momenta of the form:

$$\exp(-ak^2) \tag{6-A.2}$$

where a is a constant. Two-Tier Quantum Theory is required for Perturbation Theory calculations in higher dimensions.

PseudoQuantum Field Theory is also needed in the HyperCosmos. It is needed for proper quantization in arbitrary coordinate systems that might be relevant in higher dimension HyperCosmos spaces. For example, consider coordinate systems for non-static space-times where no time-like Killing vector exists.[32]

It also enables canonical higher order derivative theories for quark confinement and other purposes. And it dovetails with Two-Tier Coordinates to "dress" bare fermion and boson particles.

6-A.1 Reasons For Two-Tier Quantum Theory

Originally Two-Tier coordinates were developed by this author to remove infinities that appear in perturbation theory calculations. We showed that the quantum smeared coordinates of Two-Tier Quantum Field Theory succeeded in removing all ultra-violet infinities in perturbation theory including the fermion triangle infinities.

Remarkably the high precision, "low" energy[33] predictions of QED remained true in Two-Tier QED and thus remained consistent with experiment to a hitherto unsurpassed level of accuracy. "Low" energy predictions in other quantum field theories also remained unchanged. At high energies, Two-Tier perturbation theory results are finite and consequently all ultra-violet infinities, to any order in perturbation theory, in *any number of space-time dimensions* were eliminated.

In addition to removing perturbation theory infinities, Two-Tier coordinates enable us to define finite theories of Quantum Gravity and 'non-renormalizable' quantum field theories based on polynomial Lagrangians, to tame vacuum fluctuations, to eliminate infinities associated with the Big Bang, and possibly to generate the explosive growth of the universe in its role as a type of Dark Energy.[34]

6-A.2 Two-Tier Features in 4-Dimensional Space-Time

Two-Tier Quantum Field Theory[35] is based on a new method[36] in the Calculus of Variations that uses two 'layers' of fields to introduce quantum coordinates. We shall consider this technique for the specific case of a massless vector field $Y^\mu(y)$ where the index μ ranges from 0 through 3. It is analogous to the electromagnetic field.

---

[32] B. DeWitt, Phys. Rep. **19**, 295 (1975) and references therein. S. Blaha, "New Framework for Gauge Field Theories", IL Nuovo Cimento **49A**, 113 (1979).

[33] Relative to a mass scale that was perhaps of the order of the Planck mass.

[34] See Blaha (2017b) and earlier books for details. This section is basically a summary of some features.

[35] See Blaha (2005a), and Blaha (2002), for discussions of this new method to eliminate infinities in quantum field theory calculations.

[36] See Blaha (2005a)..

The X^μ coordinate system, where it appears, has a c-number real part and a q-number imaginary part. Thus particle fields which are normally defined on real four-dimensional real space-time will now be defined on a "slightly" complex four-dimensional space-time:

$$X^\mu(y) = y^\mu + i\, Y^\mu(y)/M_c^2 \tag{6-A.3}$$

where M_c is an extremely large mass of the order of the Planck mass or larger.

The $Y^\mu(y)$ field is a function of the space y coordinates. The real part of the space-time dimensions will be taken to be the space of real-valued y coordinates.[37]

The imaginary part of space-time coordinates is the massless $Y^\mu(y)$ vector quantum field that is suppressed by the very large mass scale. The effects of Quantum Dimensions only become appreciable in quantum field theory at energies of the order of M_c. At these energies exponential Gaussian factors in each particle (and ghost) propagator are generated by the Quantum Dimensions and serve to make *all* perturbation theory calculations ultra-violet finite – including calculations in Quantum Gravity. Later we will see that the Two-Tier formalism may be extended directly to the universes of the HyperCosmos with similar results – finiteness in Perturbation Theory.

The Two-Tier formalism introduces a new form of interaction that does not have the form of the simple polynomial interactions that have hitherto dominated quantum field theories. This form of interaction takes place via the composition of quantum fields and can be called a *Dimensional Interaction* or an *Interdimensional Interaction* since it affects particle behavior through Quantum Dimensions.

The basic *ansatz* of the Two-Tier formalism is to replace every appearance of a coordinate x in a quantum field with the variable

$$x^\mu \rightarrow X^\mu = (y^0, \mathbf{y} + \mathbf{Y}(y^0, \mathbf{y})/M_c^2) \tag{6-A.4}$$

where $\mathbf{Y}(y^0, \mathbf{y})$ is the spatial part of a free massless vector field with features that are identical to the free QED field in the Radiation gauge.

Then one finds that the momentum space free field Feynman propagators G(k) of all particles acquires a Gaussian factor exp(h(k)):

$$G(k) \rightarrow G(k) \exp(h(k)) \tag{6-A.5}$$

so that all perturbation theory diagrams are finite. The result is finite perturbative results for all calculations to any order in perturbation theory. Blaha (2005a) shows that Two-Tier theories are finite, Poincare covariant, and unitary. (See Blaha (2005a) for a complete discussion.)

---

[37] In a deeper theory the real part might also be a quantum field that undergoes a condensation to generate c-number coordinates. We will not consider this possibility in this book.

6-A.3 Two-Tier Quantum Coordinates Formalism

In this section we will introduce the basic Two-Tier formalism. Taking the Lagrangian described in Blaha (2005a):

$$\mathscr{L}(y) = \mathscr{L}_F(X^\mu(y))J + \mathscr{L}_C(X^\mu(y), \partial X^\mu(y)/\partial y^\nu, y) \qquad (6\text{-A.6})$$

where

$$X^\mu(y) = y^\mu + i\, Y^\mu(y)/M_c^2 \qquad (6\text{-A.7})$$

with M_c being a large mass scale, $Y_\mu(y)$ a vector quantum field, and where J is the absolute value of the Jacobian of the transformation from X to y coordinates:

$$J = |\partial(X)/\partial(y)|$$

The Lagrangian term $\mathscr{L}_C$ is

$$\mathscr{L}_C = +\tfrac{1}{4} M_c^4 F^{\mu\nu}F_{\mu\nu}$$

with

$$\begin{aligned} F_{\mu\nu} &= \partial X_\mu/\partial y^\nu - \partial X_\nu/\partial y^\mu \\ &\equiv i\,(\partial Y_\mu/\partial y^\nu - \partial Y_\nu/\partial y^\mu)/M_c^2 \end{aligned} \qquad (6\text{-A.8})$$

The Lagrangian term $\mathscr{L}_F(X^\mu(y))$ contains the terms for scalar, fermion and other gauge terms in general. The sign in $\mathscr{L}_C$ is not negative – contrary to the conventional electromagnetic Lagrangian. The reason for this difference is that the quantum field part of X^μ is imaginary. Thus $\mathscr{L}_C$ ends up having the correct sign after taking account of the factor of i in the field strength $F_{\mu\nu}$.

Defining

$$F_{Y\mu\nu} = (\partial Y_\mu/\partial y^\nu - \partial Y_\nu/\partial y^\mu)$$

we see the Lagrangian assumes the form of the conventional electromagnetic Lagrangian:

$$\mathscr{L}_C = -\tfrac{1}{4} F_Y^{\mu\nu}F_{Y\mu\nu}$$

The action of this theory has the form

$$I = \int d^4y\, \mathscr{L}(y)$$

Since $X^\mu(y)$ has an imaginary part there would appear to be an issue with the conservation of probability (unitarity). *We show unitarity is not a problem later in this appendix in section 6-A.8.*

6-A.4 Y^μ Gauge

The gauge invariance of the Lagrangian allows us to choose a convenient gauge. The gauge invariance of the full Lagrangian:

$$\mathscr{L}_S = L_F(\phi(X), \partial\phi/\partial X^\mu)\, J + \mathscr{L}_C(X^\mu(y), \partial X^\mu(y)/\partial y^\nu)$$

is based on the standard gauge invariance of $\mathscr{L}_C$, and the gauge invariance of $J\mathscr{L}_F$ in the form of translational invariance

$$X^\mu(y) \rightarrow X^\mu(y) + \delta X^\mu(y)$$

for the special case of a translation of X with the form of a gauge transformation:

$$\delta X^\mu(y) = \partial\Lambda(y)/\partial y_\mu$$

In this case we find

$$\int d^4y\, \Lambda(y)\, \partial\, [\, J\, \partial/\partial X^\mu\, \mathscr{T}_{F\mu\nu}\,]/\partial y_\nu \;=0 \qquad (6\text{-A.9})$$

after a partial integration. Thus we have the differential conservation law:

$$\partial\, [\, J\, \partial \mathscr{T}_{F\mu\nu}/\partial X^\mu]/\partial y_\nu \;=0$$

since $\Lambda(y)$ is arbitrary. This conservation law is trivially obeyed:

$$\partial \mathscr{T}_{F\mu\nu}/\partial X^\mu = 0 \qquad (6\text{-A.10})$$

Thus translational invariance in the $\mathscr{L}_F$ sector together with standard gauge invariance in the $\mathscr{L}_C$ sector automatically guarantees Y field gauge invariance of the total Lagrangian. We use the separate invariance of each term of

$$L = \int d^4y\, [\mathscr{L}_F\, J + \mathscr{L}_C\,] = \int d^4X\, \mathscr{L}_F + \int d^4y\, \mathscr{L}_C = L_F + L_C$$

under a constant translation $X^\mu \rightarrow X^\mu + \delta X^\mu$ where δX^μ is constant. Then we consider a position dependent translation/gauge transformation, which taken together with the above equation, establishes the invariance under the position dependent translation/gauge transformation.

An alternate approach that leads to the same result is to start with the particle part of the Lagrangian $\mathscr{L}_F$ rewritten to be invariant under general coordinate transformations, as it must, when we generalize to include General Relativity. Since position dependent translations are a form of general coordinate transformation the full theory must be invariant under position dependent translations due to invariance under general coordinate transformations.

Having established invariance under gauge transformations we now choose to use the most convenient gauge – the radiation gauge[38]:

$$\partial Y^i/\partial y^i = 0 \tag{6-A.11}$$

where i = 1, 2, 3. In the absence of external sources, we set

$$Y^0 = 0$$

since Y^0 does not have a canonically conjugate momentum. A conventional treatment leads to the equal time commutation relations:

$$[Y^\mu(\mathbf{y}, y^0), Y^\nu(\mathbf{y}', y^0)] = [\pi^\mu(\mathbf{y}, y^0), \pi^\nu(\mathbf{y}', y^0)] = 0 \tag{6-A.12}$$
$$[\pi_j(\mathbf{y}, y^0), Y_k(\mathbf{y}', y^0)] = -i\, \delta^{tr}{}_{jk}(\mathbf{y} - \mathbf{y}')$$

where (Note the locations of the j indexes above introduce a minus sign.)

$$\pi^k = \partial \mathscr{L}/\partial Y_k'$$
$$\pi^0 = 0$$

$$\delta^{tr}{}_{jk}(\mathbf{y} - \mathbf{y}') = \int d^3k\, e^{i\,\mathbf{k}\cdot(\mathbf{y} - \mathbf{y}')}(\delta_{jk} - k_j k_k/\mathbf{k}^2)/(2\pi)^3 \tag{6-A.13}$$

$$Y_k' = \partial Y_k/\partial y^0$$

The Radiation gauge reveals the two degrees of freedom that are present in the vector potential. The Fourier expansion of the vector potential is:

$$Y^i(y) = \int d^3k\, N_0(k) \sum_{\lambda=1}^{2} \varepsilon^i(k, \lambda)[a(k,\lambda)\, e^{-ik\cdot y} + a^\dagger(k,\lambda)\, e^{ik\cdot y}] \tag{6-A.14}$$

where

$$N_0(k) = [(2\pi)^3 2\omega_k]^{-1/2}$$

and (since m = 0)

$$\omega_k = (\mathbf{k}^2)^{1/2} = k^0$$

with $\vec{\epsilon}(k, \lambda)$ being the polarization unit vectors for $\lambda = 1,2$ and $k^\mu k_\mu = 0$.

---

[38] It is also possible to quantize using an indefinite metric that preserves manifest Lorentz covariance as was done by Gupta and Bleuler for the electromagnetic field. We will use the Gupta-Bleuler approach later to establish covariance under special relativity later. Now we opt for manifest positivity and use the radiation gauge.

The further development of Two-Tier theory is described in Part 3 of Blaha (2005a).

6-A.5 Two-Tier Uncertainty Principle

The Uncertainty Relation for Quantum Mechanics is based on the coordinate-momentum commutator. Similarly, in defining Quantum Coordinates we have established a commutation relation based on Y^{μ}. Its conjugate momentum is

$$P^{\mu}(y) = i\pi_Y{}^{\mu}(y)/M_c{}^2 \tag{6-A.15}$$

where

$$\pi_Y{}^{\mu}(y) = - dY^{\mu}/(y)dy^0$$

In the Radiation gauge (eq. 6-A.11) we see

$$\pi_Y{}^0(y) = 0 \tag{6-A.16}$$

and

$$[X^0, P^0] = [y^0, p^0] = 0 \tag{6-A.17}$$

The non-zero equal time commutation relation expressing a quantum Uncertainty Relation is

$$[P^i(\mathbf{y}, y^0) , Y^k(\mathbf{y}', y^0)] = -[\pi_Y{}^i(y), Y^k(y),]/M_c{}^4 \tag{6-A.18}$$

$$= i\delta^{trik}(\mathbf{y} - \mathbf{y}')/M_c{}^4$$

using eq. 6-A.12.

At low energy the impact of the Two-Tier Uncertainty Relation is diminished (more or less eliminated) by the factor $M_c{}^4$. Then Two-Tier Quantum Field Theory becomes ordinary Quantum Field Theory with the same results in Perturbation Theory. This limit is described in detail in Blaha (2005a), which appears in in Appendix B.

In the "low" energy limit the conventional Heisenberg Uncertainty Condition becomes evident as shown by Heitler (1954) and others. *Conventional Quantum Mechanics is a result of Second Quantization.*

6-A.6 Quantum Mechanics vs. Quantum Field Theory vs. Two-Tier Quantum Theory

Historically, Quantum Mechanics predated Quantum Field Theory, which, in turn, predates the Two-Tier Quantum Theory developed by the author in the early 2000's.

Logically, Two-Tier Quantum Theory, which embodies an Uncertainty Relation at ultrahigh energies, is the fundamental Quantum Theory. It is the predecessor of Quantum Field Theory, which only embodies the Heisenberg Uncertainty Relation. Quantum Field Theory is flawed by high energy divergences just as atomic physics, before the Bohr atom and Quantum Mechanics, was flawed by infinities in hydrogen atom models.

Quantum Field Theory (a "low" energy theory) implies Quantum Mechanics[39] as shown in Heitler[40] as well as elsewhere.

Quantum Mechanics is often treated as a complete, self-contained theory. On occasion paradoxes and ambiguities are found in Quantum Mechanics studies. They may be resolved within Quantum Mechanics. Any, that are not so resolved, should be considered within the framework of Quantum Field Theory, which is the ultimate forum for all quantum phenomena at "low" energy.

Note the trend of quantum theories, from the earliest at "low" energy theory, Quantum Mechanics, to the highest energy quantum theory: Two-Tier Quantum Theory,

Two-Tier Quantum Field Theory is a welcome extension of Quantum Theory to the deepest levels of Quantum Theory.

6-A.7 Two-Tier Perturbation Theory

The form of Two-Tier Perturbation Theory is similar to the Perturbation Theory of conventional Quantum Field Theory. It is described in *Quantum Theory of the Third Kind.* The reduction development for the U-matrix and the S-matrix are presented there.

6-A.8 Two-Tier Unitarity

The unitarity of Two-Tier Quantum Field Theory can be viewed in the cases of "low" energy phenomena and "high" energy phenomena with the scale mass of M_c separating the cases.

In the case of "low" energy phenomena the Two-Tier S-matrix gives results identical to conventional S-matrix results. Thus there is no conflict with unitarity for Two-Tier S-matrix results at "low" energy.

At "high" energy there are two potential issues: asymptotic states containing Y quanta; and conservation of probability. The first issue is resolved in the chapter 6 discussion in Appendix B of Blaha (2021i).

The second issue leads to a requirement to renormalize S-marix probability amplitudes such that their absolute values squared sum to unity – the unitarity condition. If we let S-matrix elements have the form S_{fi} where i represents an initial state and f represents one of its final states, then

$$\Sigma_n S_{nf}{}^* S_{ni} = \delta_{fi} \qquad (6\text{-A.19})$$

expresses the conventional unitarity condition. In Two-Tier Quantum Field Theory the unitarity condition for energies much less than M_c is the same as eq. 6-A.19 since perturbation theory results are the same as conventional Quantum Field Theory. The

---

[39] Quantum Mechanics is often treated as an independent theory. This practice may be valid mathematically but it is not valid physically. For example, many phenomena in atomic theory require explanation in terms of Quantum Field Theory.

[40] Heitler (1954).

sum over intermediate states n is restricted to states whose total energy is less than M_c by energy conservation.

For initial states i with energy E_i of the order of or greater than M_c, the intermediate state total energy and the final state total energy are also E_i by energy conservation. In this case the sum in eq. 6-A.19 *may* be

$$\sum_n S_{nf}{}^* S_{ni} = g_i \delta_{fi} \tag{6-A.20}$$

where g_i is a constant term dependent on the state i. In this circumstance unitarity may be recovered by redefining S_{ni} and S_{nf} as

$$S_{ni}' = S_{ni}/\sqrt{g_i} \tag{6-A.21}$$
$$S_{nf}' = S_{nf}/\sqrt{g_f} \tag{6-A.22}$$

Then the *normalized* Two-Tier unitarity condition can be expressed as

$$\sum_n S_{nf}'{}^* S_{ni}' = \delta_{fi} \tag{6-A.23}$$

with the understanding that $g_i = 1$ for initial states with energies much less than M_c.

Thus Two-Tier Quantum Field Theory satisfies an enhanced unitarity condition.

6-A.8.1 Two-Tier Analyticity

Two-Tier Quantum Theory has an apparent issue with analyticity. It is resolved by noting that lower energy perturbation theory results in Two-Tier calculations are the same as conventional quantum theory although the high energy limits are not divergent in Two-Tier calculations.

At high energy Two-Tier results differ from conventional results. This circumstance might be an issue but there are no experimental results on analyticity at high energy. So the point is mute—especially considering that HyperCosmos perturbation calculations for large space-time dimensions can currently only be done using Two-Tier quantum theory.

6-A.9 Two-Tier Quantum Space Theory

Quantum Space Theory is the theory of particles containing spaces internally. It was recently developed by the author.[41] This theory can directly put in the form of Two-Tier Quantum Space Theory by following the procedure described in section 6-A.2. The structure of the Octonion Spaces Spectrum is not changed by this formulation.

6-A.10 CQ Mechanics – A Union of Classical and Quantum Mechanics

The author developed a larger theory, CQ Mechanics, containing both Quantum Mechanics and Classical Mechanics that is described in Blaha (2016f). In this theory a phenomenon can be described as classical in one limit and as quantum in another limit.

---

[41] Blaha (2021f) and (2021g).

The description is determined by an angle that specifies one, or the other, limiting case. Intermediate cases combining both quantum and classical are also present.

This theory can bridge the classical and quantum regimes. We discuss some applications in Blaha (20016f): a generalized Feynman path integral formalism, a generalized Schrödinger equation, a generalized Boltzmann equation, the Fokker-Planck equation, a generalized approach to quantum and classical chaos, and to quantum entanglement as well as semi-quantum entanglement. Our formalism applies to Quantum Mechanics as well as the path integrals, the Fokker-Planck equation and the Boltzmann equation.

This "Mechanics" theory has an analogue at the field theoretic level, PseudoQuantum Field Theory that is briefly described next.

6-A.11 PseudoQuantum Field Theory

PseudoQuantum Field Theory was developed by the author in the 1970's and presented in a series of articles including the papers below.

S. Blaha, "The Local Definition of Asymptotic Particle States", IL Nuovo Cimento **49A**, 35 (1979). It describes the PseudoQuantization of boson and fermion field theories for use in the quantization of fields in universes and the Megaverse.

S. Blaha, "New Framework for Gauge Field Theories", IL Nuovo Cimento **49A**, 113 (1979). It describes the PseudoQuantization of gauge field theories for the purposes of defining higher derivative field theories and for use in the quantization of fields in universes and the Megaverse.

PseudoQuantum Field Theory (and its Quantum Mechanics analogue CQ Mechanics[42]) originated in the need to second quantize in unusual coordinate systems, and in curved space-time coordinate systems. It is also very relevant for the canonical formulation of higher derivative field theories for quark confinement and other applications.

PseudoQuantum Field Theory is formulated by duplicating all fields, both fermion and boson fields, in a "normal" Lagrangian theory. Scalar field theory provides a simple example that illustrates the PseudoQuantum Field Theory procedure.

We duplicate a scalar field generating two scalar fields: $\varphi_1(x)$ and $\varphi_2(x)$. We choose $\varphi_1(x)$ to have a zero equal time commutator with $d\varphi_1(x)/dx^0$ and $\varphi_2(x)$ to have a conventional equal time commutator with $d\varphi_2(x)/dx^0$. Conceptually we view $\varphi_1(x)$ is a "classical" field and $\varphi_2(x)$ is a quantum field. A Lagrangian that implements these choices of commutation relations is:

$$\mathcal{L} = \partial^\mu \varphi_1(x)\partial_\mu\varphi_2(x) - \tfrac{1}{2}\,\partial^\mu \varphi_1(x)\partial_\mu\varphi_1(x) - m_2^{\,2}\,\varphi_1(x)\varphi_2(x) + \tfrac{1}{2}\,m_1^{\,2}\,\varphi_1(x)^2 \quad (6\text{-A.24})$$

$$(\Box + m_2^{\,2})\varphi_1(x) = 0 \quad (6\text{-A.25})$$

$$(\Box + m_2^{\,2})\varphi_2(x) - (\Box + m_1^{\,2})\varphi_1(x) = 0 \quad (6\text{-A.26})$$

---

[42] See Blaha (2016f) for details, which contains Blaha (2016f). CQ Mechanics encompasses both classical mechanics and quantum mechanics, and provides a method of rotating between them. It has applications to transitions between Quantum/Semi-Classical Entanglement, and Quantum/Classical Path Integrals, and Quantum/Classical Chaos.

The canonical momenta are

$$\pi_1 = d\varphi_2(x)/dt - d\varphi_1(x)/dt \tag{6-A.27}$$
$$\pi_2 = d\varphi_1(x)/dt \tag{6-A.28}$$

and the equal time commutation relations are

$$[\varphi_i(x), \pi_j(y)] = i\delta_{ij}\delta(\mathbf{x} - \mathbf{y}) \tag{6-A.29}$$
$$[\varphi_i(x), \varphi_j(y)] = [\pi_i(x), \pi_j(y)] = 0 \tag{6-A.30}$$

implying

$$[\varphi_1(x), d\varphi_1(y)/dt] = 0 \tag{6-A.31}$$
$$[\varphi_2(x) , d\varphi_2(y)/dt] = i\delta(\mathbf{x} - \mathbf{y}) \tag{6-A.32}$$
$$[\varphi_1(x) , d\varphi_2(y)/dt] = i\delta(\mathbf{x} - \mathbf{y}) \tag{6-A.33}$$

6-A.11.1 Color Confinement

Chapter 13 discusses a non-Abelian gauge field theory that gives explicit color confinement using a PseudoQuantum formulation that leads to a confining r-potential.

If we set $m_2 = 0$ in the above example, we see a fourth order field equation results

$$\Box^2\varphi_2(x) = 0 \tag{6-A.34}$$

Color confinement results in the non-Abelian gauge field case in a similar manner.

6-A.11.2 PseudoQuantum Quantization for Non-Static Coordinate Systems

The PseudoQuantum formalism was developed by the author for "normal" and non-static coordinate systems in the 1970s.

6-A.11.3 Advantages of PseudoQuantum Quantization

In this section we point out some of its advantages in a variety of field theory contexts that are relevant for the HyperCosmos and Quantum Field Theory in general.

Some advantages of PseudoQuantum Field Theory are:

4. Quantization in any coordinate system in flat or curved space-times with an invariant definition of asymptotic particle states. *This is especially important for the higher dimension spaces of the HyperCosmos.* An n particle asymptotic state in one coordinate system is a unitarily equivalent n particle asymptotic state in any other coordinate system. Therefore particle number is invariant under change of coordinate system. This is important for the Unified SuperStandard Theory in curved space-times. It is also important for quantization in higher dimensional spaces such as those of the HyperCosmos. The method was developed in the late 1970's by the author to provide a quantization procedure which supports a unique particle interpretation of states in arbitrary non-static space-times where no global time-like coordinate (Killing vector) exists.

PseudoQuantum Field Theory which we developed in a series of books[43] also can be formulated in the Octonion Spectrum of spaces. For example, we can use it to implement the Higgs Mechanism to generate particle masses and symmetry breaking.

5. PseudoQuantum Field Theory enables one to define Higgs particle dynamics in such a way that a non-zero vacuum expectation value cleanly separates from the quantum field part of the Higgs fields. This technique can be used in symmetry breaking mechanisms, mass generation, and possible generation of coupling constants as vacuum expectation values.

6. It supports the canonical definition of higher derivative field theories through the use of the Ostrogradski bootstrap. We have used it to construct a fourth order theory of the Strong interaction that has color confinement and a linear r potential and a higher derivative theory of Gravity.. (See chapters 13 and 14, and Blaha (2017b) for details.)

6-A.12 Combination of Two-Tier and PseudoQuantum Formalisms

These formalisms can be directly combined with no issues.

6-A.13 A Model Illustrating Scalar Field Quantization Using X^μ

We begin by considering the case of a scalar quantum field theory. We assume a real underlying y subspace. Since X^μ is a set of coordinates, we choose to define a scalar field ϕ as a function of X^μ, which, in turn, is a function of the y^ν coordinates. We will provisionally second quantize ϕ treating X^μ as c-number coordinates using a conventional approach.[44]

We assume a Lagrangian, with the momentum conjugate to ϕ:

$$\pi_\phi = \partial L_F / \partial \phi' \equiv \partial L_F / \partial(\partial\phi/\partial X^0) \qquad (6\text{-A.35})$$

Following the canonical quantization procedure, π and ϕ become Hermitian operators with equal time ($X^0 = X^{0\prime}$) commutation rules:

$$[\phi(X), \phi(X')] = [\pi_\phi(X), \pi_\phi(X')] = 0 \qquad (6\text{-A.36})$$
$$[\pi_\phi(X), \phi(X')] = -i\,\delta^3(\mathbf{X} - \mathbf{X'})$$

The standard Fourier expansion of the solution to the Klein-Gordon equation is:

---

[43] See Blaha (2017b) for the discussion of the PseudoQuantum field theory formalism for Higgs particles in our Extended Standard Model. See chapter 20 of Blaha (2017b), and earlier books, for a more detailed view than that presented here.

[44] Some texts are: Bogoliubov, N. N., Shirkov, D. V., *Introduction to the Theory of Quantized Fields* (Wiley-Interscience Publishers Inc., New York, 1959); Bjorken, J. D., Drell, S. D., *Relativistic Quantum Fields* (McGraw-Hill, New York, 1965); Huang, K., *Quarks, Leptons & Gauge Fields Second Edition* (World Scientific, River Edge, NJ, 1992); Kaku, M., *Quantum Field Theory* (Oxford University Press, New York, 1993); Weinberg, S., *The Quantum Theory of Fields* (Cambridge University Press, New York, 1995).

$$\phi(X) = \int d^3p\, N_m(p)\, [a(p)\, e^{-ip\cdot X} + a^\dagger(p)\, e^{ip\cdot X}] \qquad (6\text{-A.36a})$$
$$= \int d^{3k}p\, N_m(p)\, [a(p) \exp(-ip\cdot(y + Y/M_c^2)) + $$
$$+ a^\dagger(p) \exp(ip\cdot(y + Y/M_c^2))] \qquad (6\text{-A.36b})$$

where

$$N_m(p) = [(2\pi)^3 2\omega_p]^{-1/2}$$

and

$$\omega_p = (\mathbf{p}^2 + m^2)^{1/2}$$

The commutation relations of the Fourier coefficient operators are:

$$[a(p), a^\dagger(p')] = \delta^3(\mathbf{p} - \mathbf{p'})$$
$$[a^\dagger(p), a^\dagger(p')] = [a(p), a(p')] = 0$$

The reader will recognize the quantization procedure is formally identical to the standard canonical quantization procedure of a free scalar quantum field.

In the case of spin ½, spin 1 and spin 2 fields the standard quantization procedure *in terms of the X coordinate system* can also be followed in a way similar to the procedure in standard texts.

6-A.14 Scalar Feynman Propagators

The momentum space free field Feynman propagators G…(k) of all particles and ghosts in all Two-Tier Quantum Field Theories acquires a Gaussian factor exp(h(k)):

$$G\ldots(k) \rightarrow G\ldots(k) \exp(h(k))$$

so that all perturbation theory diagrams are finite. The result is a finite perturbation in all calculations to any order in perturbation theory. Blaha (2005a) shows that Two-Tier theories are finite, Poincare covariant, and unitary.

7. The HyperCosmos and Our N = 7 Universe

Fig. 1.1 shows the HyperCosmos Blaha space number N = 7 has a four dimension space-time. Consequently we view it as the space of our universe. It has a dimension array of $d_{d7} = 256$ dimensions. We separate this array into a 16×16 square array (Fig. 7.1) for the purpose of mapping its dimensions to the spectrum of fundamental fermions and to the symmetry groups of the Unified SuperStandard Theory (UST).[45]

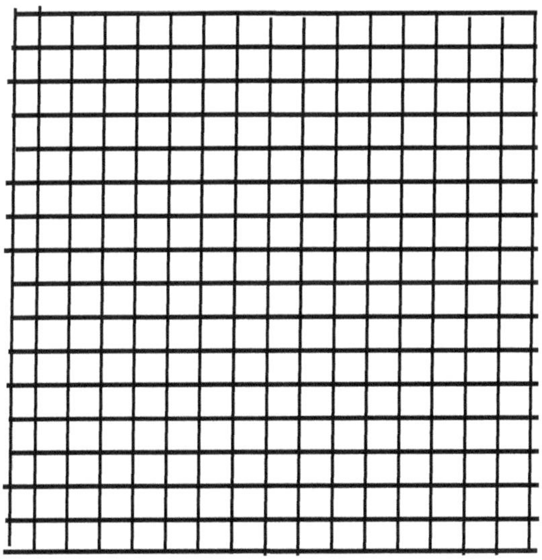

Figure 7.1. 16×16 dimension array.

7.1 Fundamental Fermion Spectrum

The square form of the dimension array lends itself naturally to consisting of four layers, each of four generations for both Normal and Dark sectors with a view to Normal – Dark symmetry. Figs. 7.2 and 7.3 show the fundamental fermion spectrum with each block containing four generations and with fermion species arranged in SU(4)-plets (that could be easily brought to the form of quark SU(3)-plets and leptons.

This fermion spectrum was first anticipated in the UST. See for example Blaha (2018e). It led the author to pursue hypercomplex number-based theories in the past three years.

[4545] This was first presented as the QUeST model in January, 2020. See Blaha (2020a).

SU(4)- Based Fermion Particle Periodic Table

Number of Columns =	4 NORMAL	4	4 DARK	4
Layer 1 4	e 3 up-quarks	ν 3 down-quarks	e 3 up-quarks	ν 3 down-quarks
Layer 2 4	e 3 up-quarks	ν 3 down-quarks	e 3 up-quarks	ν 3 down-quarks
Layer 3 4	e 3 up-quarks	ν 3 down-quarks	e 3 up-quarks	ν 3 down-quarks
Layer 4 4	e 3 up-quarks	ν 3 down-quarks	e 3 up-quarks	ν 3 down-quarks

Figure 7.2. The QUeST 16 $\times$ 16 fermion spectrum of our universe tentatively arranged as SU(4)-plets that correspond directly with SU(4) (or SU(3)$\otimes$U(1)) fermions. There are four layers. Each set of 4 fermions has 4 generations matching the number of rows in each layer. This Periodic Table is broken into Normal and Dark sectors.

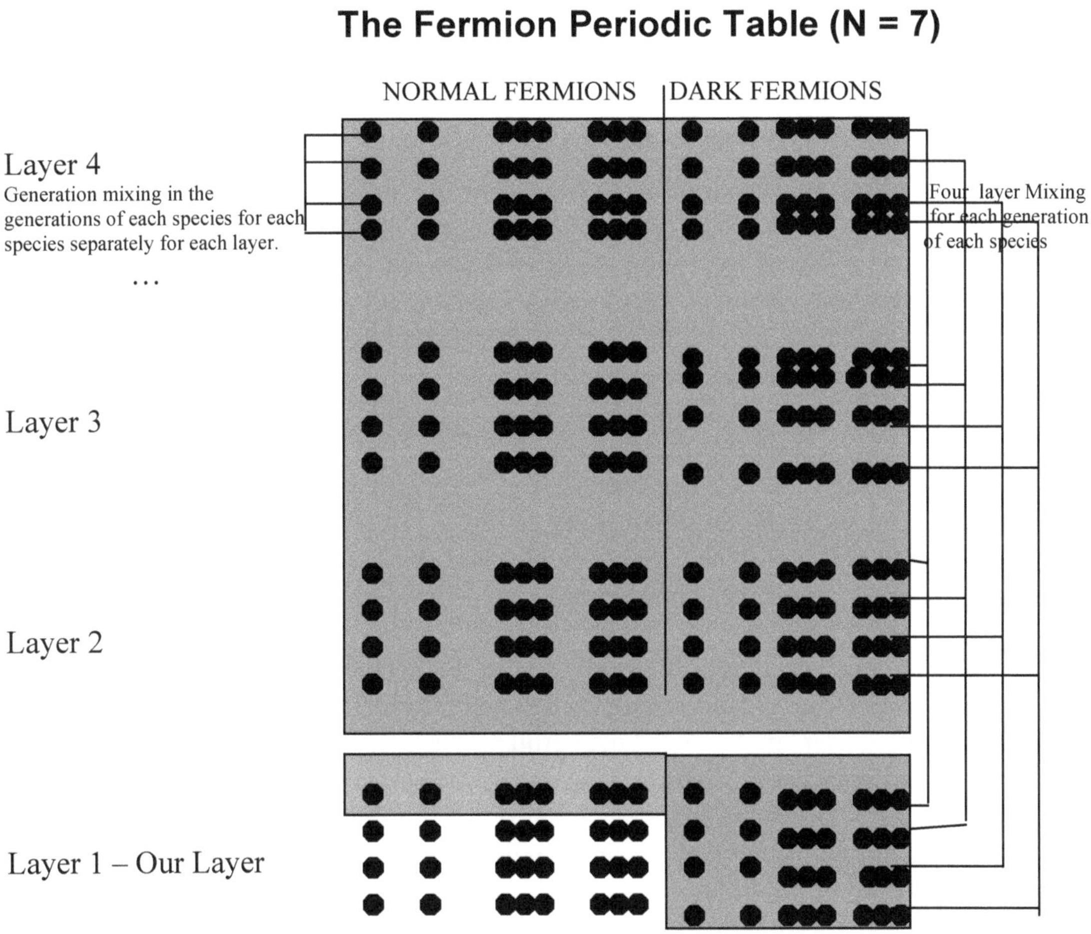

Figure 7.3. QUeST Fermion particle spectrum and partial examples of the pattern of mass mixing of the Generation groups and of the Layer groups. Unshaded parts are the known fermions, A shaded generation is an additional, as yet not found, 4th generation. The lines on the left side (only shown for one layer) display the Generation Group mixing within each layer. The Generation mixing occurs within each layer using a separate Generation group for each layer. The lines on the right side show Layer group mixing (for Dark matter) with the mixing among all four layers for each of the four generations individually. There are four Layer groups for Normal matter and four Layer groups for Dark matter. There are 256 fundamental fermions. QUeST and UST have the same fermion spectrum.

7.2 The Symmetry Groups for N = 7 Space, QUeST and UST

In January 2020, the author showed the QUeST hypercomplex model had the same symmetry group structure as the UST which he developed in prior years. The 256 component dimension array d_{d7} (which QUeST had) is directly mapped to the set of irreducible group representation dimensions of the UST.

The total of the fundamental representations' dimensions equals the 256 dimensions of the N = 7 dimension array. Seven of the SL(2,**C**) groups are mapped to Connection Groups for good reason in Appendix 7-B.

7.3 Groups Specified by Irreducible Representations Dimensions

The map of the N = 7 HyperCosmos space (QUeST) dimensions to UST symmetry groups is based on matching dimensions in the dimension array to the irreducible representation dimensions of various unitary simpler groups.

U(1) – 2 real dimensions
SU(2) – 4 real dimensions
SU(2)⊗U(1) – 4 real dimensions
SU3) – 6 real dimensions
U(4) – 8 real dimensions
SU(4) – 8 real dimensions
SL(2,C)[46] – 4 real dimensions

The matching procedure is clearly not unique. However the process becomes unique if one requires the groups are generated as replicas from the N = 7 Fundamental Reference Frame (chapter 8) with the groups specified in it determined by the considerations in chapter 9.

The first step (chapter 8) is the generation of the Fundamental Reference Frame (FRF) from a combined General Relativistic – Internal Symmetry transformation. For N = 7 the FRF has a set of 16 dimensions which are used to specify a primary fundamental set of fermions and also a primary set of symmetries. The 16 dimensions transform to 16 replicas of 16 dimensions. Likewise the primary fermions replicate to 256 fundamental fermions and the 16 symmetry dimensions replicate to the 256 symmetry dimensions of QUeST (UST).

Secondarily, for N = 7, the 16 FRF dimensions represent 16 primary fundamental fermions that are separated into sets of 4 fermions by the space-time subluminal-superluminal differentiation due to two time coordinates. (Chapter 9) The 16 FRF dimensions also represent 16 symmetry group fundamental representation dimensions that separate into sets of 4 fundamental representation dimensions by space-time subluminal-superluminal differentiation due to two time coordinates. These sets are U(4)⊗U(4) initially. They replicate 16 times due to the transformation. They may then "break" after transformation to

---

[46] The Lorentz group. $SO^{+}(1,3)$ in particular

SU(4)⊗SU(2)⊗U(1)⊗SL(2,C) or
SU(3)⊗U(1)⊗SU(2)⊗U(1)⊗SL(2,C) or
SU(4)⊗SU(2)⊗U(1)⊗U(2) or
SU(3)⊗U(1)⊗SU(2)⊗U(1)⊗U(2)

where the last U(2) group factor is a Connection. Group (See Appendix 7-B.)

Thus the matching procedure is unique if the FRF transformation is taken into account.

7.4 The N = 7 QUeST Features in the UST

Other important features of the UST are specified in the following chapters.

1. Generation and Layer Groups – Appendices 7-A and 7-B.
2. Gauge vector bosons – follow from the specification of groups. Chapter 9.
3. Higgs particles – also follow from the group specification. Chapter 9
4. Masses – due to Higgs Mechanism.
5. Interactions – specified by minimal gauge coupling considerations.[47] Chapter 12.

7.5 The Standard Model

QUeST and UST contain the Standard Model of Elementary Particles. See Chapter 11.

[47] We modify the Strong and Gravitation interactions. See chapter 12.

Number of Columns =	4 NORMAL	4	4 DARK	4
Layer 1 4	SU(4)⊗ Generation U(4)	SU(2)⊗U(1)⊗SL(2,C)[48] Layer U(4)	SU(4)⊗ Generation U(4)	SU(2)⊗U(1)⊗SL(2,C) Layer U(4)
Layer 2 4	SU(4)⊗ Generation U(4)	SU(2)⊗U(1)⊗SL(2,C) Layer U(4)	SU(4)⊗ Generation U(4)	SU(2)⊗U(1)⊗SL(2,C) Layer U(4)
Layer 3 4	SU(4)⊗ Generation U(4)	SU(2)⊗U(1)⊗SL(2,C) Layer U(4)	SU(4)⊗ Generation U(4)	SU(2)⊗U(1)⊗SL(2,C) Layer U(4)
Layer 4 4	SU(4)⊗ Generation U(4)	SU(2)⊗U(1)⊗SL(2,C) Layer U(4)	SU(4)⊗ Generation U(4)	SU(2)⊗U(1)⊗SL(2,C) Layer U(4)

Figure 7.4. The SU(4) based pattern of symmetries of QUeST and UST. Note each layer has its own set of internal symmetries. Normal and Dark sector symmetries are also different. The Generation Groups, theLayer Groups and the Connection Groups (not shown here) tie the various parts together. The origin and role of these groups is described in Appendices 7-A and 7-B. The pattern of symmetries matches those of QUeST. SU(4) is broken to SU(3)⊗U(1) experimentally.The eight SL(2, **C**) groups are transformed into one 4-dimension space-time and seven U(2) Connection Groups in Appendix 7-B.

NORMAL		DARK	
SU(3)⊗U(1) Generation U(4)	SU(2)⊗U(1)⊗SL(2, C) Layer U(4)	SU(3)⊗U(1) Generation U(4)	SU(2)⊗U(1)⊗SL(2, C) Layer U(4)
SU(3)⊗U(1) Generation U(4)	SU(2)⊗U(1)⊗SL(2, C) Layer U(4)	SU(3)⊗U(1) Generation U(4)	SU(2)⊗U(1)⊗SL(2, C) Layer U(4)
SU(3)⊗U(1) Generation U(4)	SU(2)⊗U(1)⊗SL(2, C) Layer U(4)	SU(3)⊗U(1) Generation U(4)	SU(2)⊗U(1)⊗SL(2, C) Layer U(4)
SU(3)⊗U(1) Generation U(4)	SU(2)⊗U(1)⊗SL(2, C) Layer U(4)	SU(3)⊗U(1) Generation U(4)	SU(2)⊗U(1)⊗SL(2, C) Layer U(4)

Figure 7.5. The SU(3)⊗U(1) form of the 4 layers of symmetry groups of an N = 7 universe of the form in QUeST and UST. Seven of the SL(2, **C**) groups are transformed to U(2) Connection groups later. See Appendix 7-B.

48 The Lorentz group. $SO^{+}(1,3)$ in particular

Appendix 7-A. Generation and Layer Groups

7-A.1 U(4) Generation Groups

In the Big Bang all particles were massless and all symmetries unbroken. Hence the four Normal particle number symmetries, and the four Dark particle number symmetries, are all "conserved" in the Big Bang. Afterwards conservation laws are then broken.

We define two particle number operators for normal up-quark particles and down-quark particles, B_{uq} and B_{dq}. Similarly we define two particle number operators for normal species "e" (electron) particles and species "ν" particles, B_e and $B_ν$. Similarly we define Dark matter equivalents:[49] B_{De}, $B_{Dν}$, B_{Duq}, and B_{Ddq}.

In the absence of symmetry breaking these fermion particle number operators would be conserved. Thus there are two sets of "diagonal" operators with associated U(4) groups for the Normal and Dark sectors. They are part of the Normal U(4) Generation Group and the Dark U(4) Generation Group.

The fermion fundamental representation of a U(4) group has four fermions. U(4) has rotations, and also interactions of the form $\overline{\Psi}\gamma\cdot B\cdot T\Psi$ where Ψ is a fermion four-vector, B is a 16 component U(4) gauge field, and T consists of 16 component 4 × 4 U(4) arrays.

In the case of the Generation Group the gauge fields have electric charge zero. Since the four species have different electric charges (1, 0, 2/3, -1/3) the U(4) gauge boson fields cannot mix the fermions of different species. Generation Group interactions are diagonal[50] in fermion species (e, ν, up-quark, and down-quark species).

Consequently the U(4) Generation Group for each layer[51] must have a reducible representation D consisting of a set of four fundamental U(4) representations, D_e, $D_ν$, D_{upq}, and D_{dnq}, appearing in blocks along the diagonal of D. Each block is a separate U(4) irreducible representation for a species (due to the electric charge superselection rule.) There is a U(4) Generation group for each of the four layers of the Normal and Dark sectors totaling to 8 generation groups.

There are four generations of each species in the Normal and in the Dark matter sectors. The four generations for each fermion species: e, ν, up-quark, and down-quark each furnish a U(4) fundamental representation within the reducible representation D. The fourth generation of normal fermions has not as yet been found due to their extremely large masses.

---

[49] By analogy, we assume that there are four species of Dark matter: charged Dark leptons, neutral Dark leptons, Dark up-type quarks, and Dark down-type quarks. Thus we are led to the Dark particle numbers: Dark Baryon Numbers, and Dark Lepton Numbers shown above.

[50] ElectroWeak interactions can cross between species due to their charged gauge vector bosons.

[51] For the Normal and Dark sectors separately.

The Generation Group rotates the fundamental fermions of each fundamental representation separately for each of the four species of each of the four layers.[52] Thus the Generation Group guarantees that all generations of each species have the same electric charge and other quantum numbers.

The U(4) Generation Group also specifies a gauge field interaction among the fermions of its fundamental representation, species by species, for both Normal and Dark sectors. The form of the interactions for the Normal sector for each fermion layer is:

$$g_e\Psi_e\gamma\cdot B_e\cdot T\Psi_e + g_v\overline{\Psi}_v\gamma\cdot B_v\cdot T\Psi_v + g_{upq}\overline{\Psi}_{upq}\gamma\cdot B_{upq}\cdot T\Psi_{upq} + g_{dnq}\overline{\Psi}_{dnq}\gamma\cdot B_{dnq}\cdot T\Psi_{dnq} \tag{7-A.1}$$

where g_e … are coupling constants, the gauge vector fields are B_e … , and the Ψ_e … are 4-vectors of fermions of the four generations of each species in a layer.

The gauge vector bosons of the Generation Group have large masses. If the conservation of the fermion particle numbers is broken then we view it as a consequence of Generation Group symmetry breaking.

Generation Group rotations guarantee the internal quantum numbers of each generation of each species are the same since symmetry breakdown is not present at the instant of the Big Bang.

The above discussion applies similarly to the Dark sector. Thus there are 8 Generation Groups in total.

7-A.2 U(4) Layer Groups

The set[53] of particle number operators can be extended if we take account of the fourfold fermion generations.

We can subdivide the above particle number sets into four additional particle numbers *per generation*. For the ith generation (of the four generations) we define

L_{ie} – The "e" species particle number for the ith generation
L_{iv} – The v species particle number for the ith generation
L_{iuq} – The up-quark species particle number for the ith generation
L_{idq} – The down-quark species particle number for the ith generation

L_{iDe} – The Dark "e" species particle number for the ith generation
L_{iDv} – The Dark v species particle number for the ith generation
L_{iDuq} – The Dark up-quark species particle number for the ith generation
L_{iDdq} – Dark down-quark species particle number for the ith generation

for each generation i = 1, 2, 3, 4. Individual fermions have positive $L_{ia} = +1$ values and antifermions have negative $L_{ia} = -1$ values for each species.

[52] There are separate Generation groups for each layer.
[53] Here again, in the Big Bang all particles were massless and all symmetries unbroken. Hence particle numbers are "conserved" in the Big Bang. Conservation is then broken afterwards in most cases.

At this point we have a set of four particle number operators for each of four generations (i = 1, 2, 3, 4) of fermions in the Normal sector and similarly in the Dark sector. We then define a U(4) group framework for each set of particle numbers.

The only way to specify fundamental representations for each of the four sets in a sector is to assume there are four layers, with each layer having four generations, and with a fundamental U(4) representation defined for each generation composed of fermions from each layer. Thus there are four Layer Groups for each Normal and each Dark sector: a Layer Group for generation 1, a Layer Group for generation 2, and so on.

The Layer Groups are also "split" by species due to the electric charge superselection rule. Each Layer Group is diagonal in the four fermion species. All their gauge fields are electrically neutral. Thus the Layer Group fundamental representations of the Normal sector total 16.[54] The Layer Group fundamental representations of the Dark sector also total 16. There are 4 Normal Layer Groups and 4 Dark Layer Groups.

Consequently each of the four U(4) Layer Groups in the Normal fermion sector has a reducible U(4) representation D_j for j = 1, 2, 3, 4. Each reducible representation is composed of four irreducible U(4) representations for each species due to the electric charge superselection rule:

$$D_j = D_{je} + D_{jv} + D_{jupq} + D_{jdnq},$$

for j = 1, 2, 3, 4.

There are four layers of each species in the Normal and in the Dark matter sectors. The second, third and fourth layers of normal fermions has not as yet been found due to their extremely large masses.

A Layer Group rotates the fundamental fermions of each fundamental representation separately for each of the four species of each of the four generations.

The Layer Groups guarantee that all layers of each species have the same electric charge and other quantum numbers.

Each U(4) Layer Group also specifies a gauge field interaction among the fermions of its fundamental representation, species by species, for both Normal and Dark sectors. The form of the interactions is:

$$g_{ei}\overline{\Psi}_{ei}\gamma\cdot C_{ei}\cdot T\Psi_{ei} + g_{vi}\overline{\Psi}_{vi}\gamma\cdot C_{vi}\cdot T\Psi_{vi} + g_{upqi}\overline{\Psi}_{upqi}\gamma\cdot C_{upqi}\cdot T\Psi_{upqi} + g_{dnqi}\overline{\Psi}_{dnqi}\gamma\cdot C_{dnqi}\cdot T\Psi_{dnqi} \tag{7-A.2}$$

for i = generation = 1, … , 4, where g_{ei} … are coupling constants, the gauge fields are C_{ei} … , and the Ψ_{ei} … are 4-vectors of fermions formed of the ith generation fermions in each layer of each species.

The gauge vector bosons of the Layer Groups also have large masses. If the conservation of the fermion particle numbers is broken then we view it as a consequence of Layer Groups symmetry breaking.

---

[54] Four Layers groups irreducible representations for the generations × four species = 16 irreducible representations.

Layer Group rotations guarantee the internal quantum numbers of each layer of each species are the same since symmetry breakdown is not present at the instant of the Big Bang.

The above discussion applies similarly to the Dark sector. There are 16 Dark Layer Groups.

Fig. 7.3 shows the fundamental fermion spectrum with the representations of the Generation groups and Layer groups indicated.

Experimentally, we know of three generations of fermions—the lowest 3 generations of the lowest level. The remaining 4th generation and three layers of fermions are of much higher mass and are yet to be found.

See Blaha (2019g) and (2018e) for a detailed discussion of the Layer Groups. We note in passing that the symmetries of these number operators are badly broken. Yet the underlying group structure remains.

7-B. Hypercomplex Space-Time Coordinates and Connection Group Symmetries

Hypercomplex numbers are known to be related to symmetry groups. In this section we consider the multiple space-time symmetries that appear in the separation of dimension arrays into representations of groups. We have suggested in Blaha (2021b), (2021e) and (2021g) that the set of hypercomplex space-time dimensions be transformed to a set symmetry groups in each HyperCosmos space.

There is a two-fold justification for this procedure: 1). There is no evidence for the existence of hypercomplex numbers in our universe. 2) The generated set of symmetry groups has the important purpose of providing ultra-weak interactions uniting Normal and Dark matter. Without unification, Dark matter becomes physically irrelevant fot elementary particle physics.

We begin by noting the space-time dimension is set by the dimension array:

$$r = \log_2 (d_{dN}/16) \tag{7-B.1}$$

For N = 9 the space-time dimension is 0 since d_{dN} = 16. For N = 7 (our universe's space) the space-time dimension is 4 since d_{dN} = 256. We will consider the cases of NEWQUeST, NEWUTMOST, and NEWMaxiverse to illustrate the method.

7-B.1 Hypercomplex Coordinates Transformed to Symmetry Groups in Our Universe N = 7

Our QUeST formulation for our N = 7 universe has a d_{d7} = 256 component dimension array. A preliminary view of the symmetry groups that it implies appears in Fig. 7-B.1. Note that there are 32 dimensions that are initially allocated to space-time dimensions suggesting a hypercomplex space-time. We chose to allocate 4 real dimensions to obtain the 4-dimension space-time implied by eq. 7-B.1. The remaining 28 real dimensions we allocated to additional symmetry groups—namely seven U(2) groups. We call these groups Connection groups since their role is to "connect" fermions residing in different fermion spectrum blocks.

The structure of the seven additional U(2) groups is not specified. We chose to use a reasonable physical principle to allocate them. We believe their role is to "connect" fermions in different blocks since the fermions within each fermion block have "known" interactions. Note that there are initially eight blocks, each with their own set of symmetry groups and corresponding interactions, and initially no interactions between the eight blocks. If there are no block interactions (except gravity), then the Physics of the fermion set is conceptually disjoint. Thus we choose to implement inter-block interactions as in Fig. 7-B.2 following the principle:

A fermion in any block has interactions either directly, or indirectly, with every other fermion in any other block.

Fig. 7-B.2 shows an implementation of this principle. The horizontal lines in Fig. 7-B.2 indicate 1:1 transformations between all corresponding fermions of each "Normal" and "Dark" block. The three "angled" lines indicate 1:1 transformations between corresponding fermions of a "Normal" and a "Dark" fermion block in the layer above it. The result is a see-saw type of pattern.

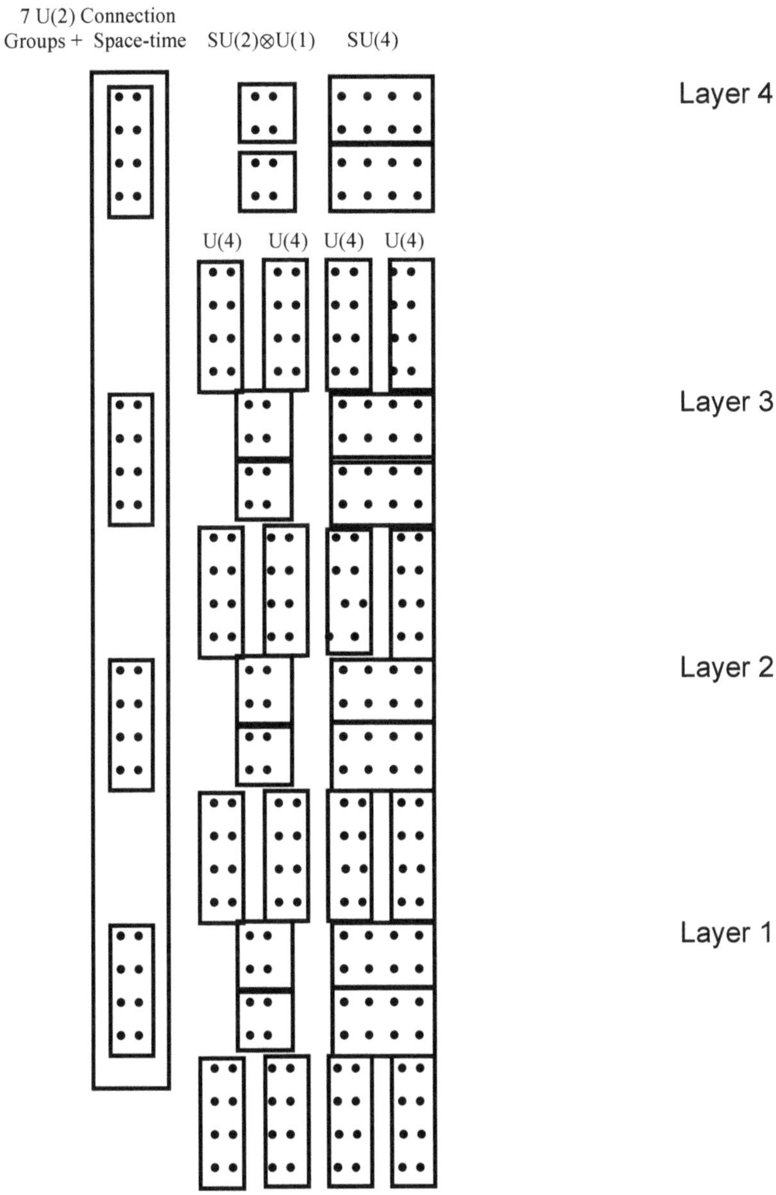

Figure 7-B.1. The four layers of QUeST, NEWUST and NEWQUeST internal symmetry groups (and space-time) with SU(4) before breakdown to SU(3)⊗U(1). Note the left column of blocks are combined below to specify a 4 dimension real space-time plus seven U(2) Connection groups. Note each layer has 64 dimensions = 56 + 8 dimensions.

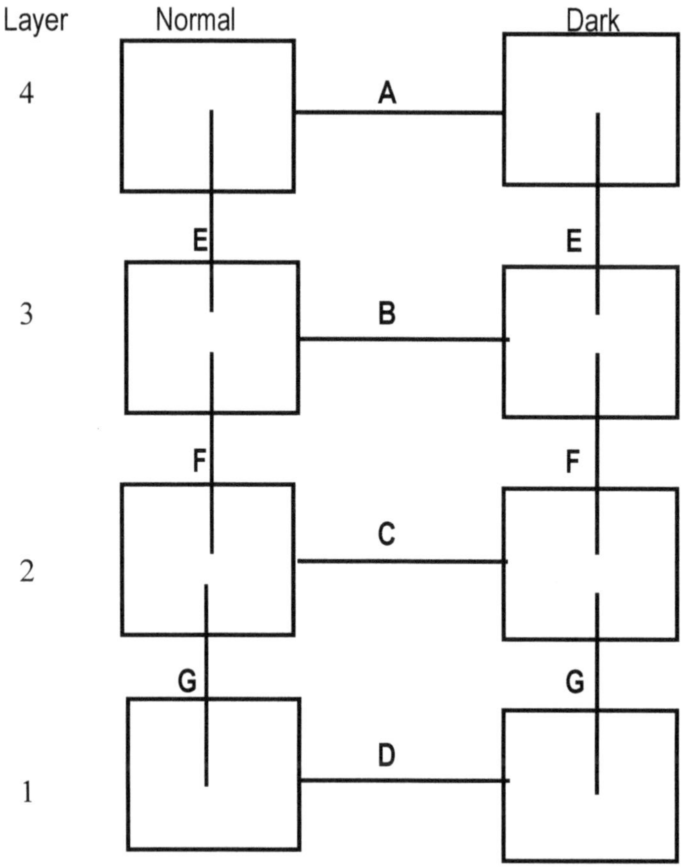

Figure 7-B.2. The seven U(2) Connection groups (shown as 10 lines) between the eight NEWQUeST/NEWUST blocks. Connection groups are obtained by transfering 28 dimensions from QUeST space-time to internal symmetries with the consequent reduction of the space-time from four octonion (complex quaternion) coordinates to four real coordinates. The Connection groups generate rotations and interactions between corresponding fermions and vector bosons of each pair of blocks. The Normal and Dark sector U(2) vertical connections above (E, F, G) represent the same U(2) groups.

7-B.1.1 The U(2) Connection Groups

The seven U(2) Connection groups of Fig. 7-B.2 generate "rotations" and interactions between *corresponding* fermions and vector bosons of each pair of blocks of the eight blocks of fermions in NEWQUeST/NEWUST.

7-B.1.1.1 Horizontal Lines

The horizontal lines in Fig. 7-B.2 (A, B, C, and D) each represent a U(2) Connection group that "rotates" two *corresponding* fermions in the Normal and Dark sectors of each layer. Thus a Normal e is "rotated" with a corresponding Dark e, and so on.

Each of the four horizontal Connection Groups has a reducible U(2) representation D that is the sum[55] of 4*8 = 32 irreducible U(2) representations. We may view the irreducible representations D_j as strung along the diagonal.

$$D = \sum_{j=1}^{32} D_j \tag{7-B.2}$$

for each of the U(2) groups of the four horizontal lines in Fig. 7-B.2.

The U(2) group also specifies gauge field interactions between corresponding fermions in each layer of the Normal and Dark sectors of the form

$$g\overline{\Psi}_{Nn}\gamma\cdot A\cdot T\Psi_{Dn} \tag{7-B.3}$$

where N indicates a Normal fermion and D indicates the corresponding Dark fermion, with A being a U(2) gauge vector boson, and n the label for corresponding fermions.

These U(2) transformations imply that the Normal and Dark sectors have the same species.

7-B.1.1.2 Vertical Lines

The pairs of vertical lines in Fig. 7-B.2 (E, F, G) each represent a U(2) Connection group that "rotates" sets of two *corresponding* fermions in adjacent layers as shown in Fig. 7-B.2 in the Normal and Dark sectors. Thus a Normal e in layer 1 is rotated with a corresponding Normal e in layer 2, and so on.

Each of the three (six counting both Normal and Dark lines in Fig. 7-B.2) vertical Connection Groups has a reducible U(2) representation D that is the sum of 64 irreducible U(2) representations.[56] We may view D as an array of 64 U(2) irreducible representation dimensions D_j strung along the diagonal.

$$D = \sum_{j=1}^{64} D_j \tag{7-B.4}$$

for each of the U(2) groups of the 3 (6) horizontal lines in Fig. 7-B.2. Note the 64 irreducible representations include both Normal and Dark sectors of a layer.[57]

Each U(2) group also specifies a gauge field interaction between corresponding fermions in adjacent layers for both Normal and Dark sectors:

$$g\overline{\Psi}_{nl_1}\gamma\cdot A\cdot T\Psi_{nl_2} \tag{7-B.5}$$

---

[55] Eight fermions per generation ▪ four generations, thus accounting for each fermion in a block.

[56] Eight fermions per generations ▪ four generations ▪ 2 types of matter (Normal and Dark).

[57] There are 64 fermions in total for each of the four layers of NEWQUeST/NEWEST.

where l_1 and l_2 designate layers, A is a gauge field vector boson, and n the label for corresponding fermions.

Each E, F, and G U(2) group reducible representation includes both Normal and Dark sectors.

7-B.1.2 The Connection Groups are UltraWeak Interactions

Since there is no convincing experimental evidence for particle interactions between Normal and Dark matter, or between Normal fermion layers the Connection groups appear to be Ultraweak.

7-B.2 NEWUTMOST with Six Real Space-Time Coordinates (Dimensions)

The NEWUTMOST space is a Blaha number N = 6 HyperCosmos space. It is the Megaverse (Multiverse) with 6 real-valued space-time coordinates by eq.7-B.1 above. It has a $d_{d6} = 1024$ dimension array.

We view the NEWUTMOST dimension array as composed of four copies of the N = 7 dimension array. Initially 4·32 = 128 dimensions are for space-time coordinates.

We allocate 4·28 = 112 dimensions to each of the four NEWQUeST copies within NEWUTMOST to give each copy 7 SU(2) Connection groups. Therefore 4* 28 = 112 of the 128 space-time dimensions of UTMOST are mapped. The remaining 16 dimensions give 6 real-valued space-time dimensions to the Megaverse space, 8 dimensions to a Megaverse SU(4) Connection group, and 2 dimensions to a U(1) group for all fermions in the Megaverse. See Figs. 7-B.3 and 7-B.4.

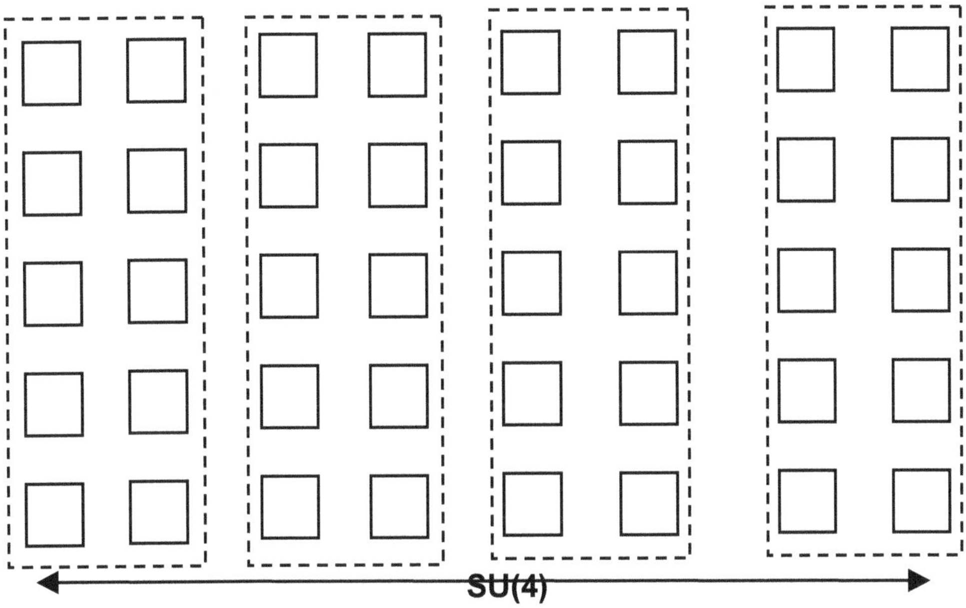

Figure 7-B.3. NEWUTMOST has four NEWQUeST copies. An SU(4) internal symmetry Connection group maps between corresponding fermions in the four copies: fermion by fermion. An additional U(1) Connection group applies to every corresponding fermion. It is not shown in this figure.

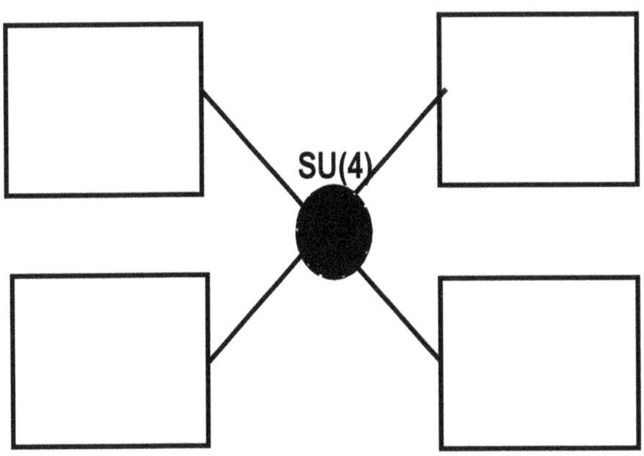

Figure 7-B.4. The SU(4) Connection Group of NEWUTMOST connecting fermions in the four NEWQUeST "copies" blocks. An additional U(1) Connection group applies to every corresponding fermion. It is not shown in this figure.

7-B.3 NEWMaxiverse with Eight Real Coordinates (Dimensions)

The NEWMaxiverse is a Blaha number N = 5 space with a dimension array of d_{d5} = 4096 dimensions.

The NEWMaxiverse contains four copies of NEWUTMOST. Each NEWUTMOST has 4*6 = 24 space-time dimensions. (The remainder in each NEWUTMOST copy consists of internal symmetries and Connection groups.) We allocate the 24 dimensions to eight real space-time dimensions plus 16 dimensions for a new ultraweak (possibly broken) SU(8) Connection group for the four parts of the NEWMaxiverse. See Fig. 7-B.5.

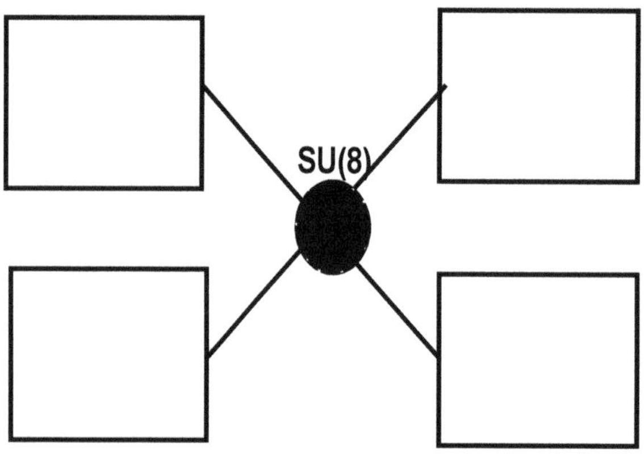

Figure 7-B.5. The SU(8) Connection Group of NEWMaxiverse connecting fermions in the four NEWUTMOST "copies" blocks.

7-B.4 Determining the Connection Groups for a Space

In the previous sections we determined the Connection Groups for N = 7, N = 6, and N = 5 spaces. The dimensions required for the Connection Groups of a space d_{gN} for $N \leq 6$ can be obtained by noting the space-time dimension of a space-time r is related to the space-time dimension r – 2 of the next lower space. Since a space is four copies of the next lower space it has 4(r – 2) space-time dimensions from the four duplicates. Then, using eq. 2.17, we find

$$4r_{N+1} = 4(r_N - 2) = r_N + d_{gN}$$

or

$$d_{gN} = 3r_N - 8 \tag{7-B.5}$$

where d_{gN} is the *real-valued* space-time dimensions reallocated to Connection Groups in space N. For N = 6, r_6 = 6, d_{gN} = 10. For N = 5, r_5 = 8, d_{gN} = 16. And so on.

Eq. 7-B.5 determines the dimensions available for Connection Groups for the space with Blaha number $N \leq 6$. Using eq. 7-B.1:

$$r_N = \log_2 (d_{dN}/16) \qquad (7\text{-B.6})$$

we relate d_{gN} to the size of the dimension array:

$$d_{gN} = 3 \log_2(d_{dN}/16) - 8 \qquad (7\text{-B.7})$$

thus specifying the number of space-time dimensions allocated to space number N.

8. Fundamental Reference Frames

A common feature of dynamical calculations in Physics is to choose a reference frame which facilitates computation. Frequent choices are a rest frame and the center of mass frame. In this chapter we find a Fundamental Reference Frame (actually an infinite set of such frames) that plays a similar role for coordinate reference frames. We describe how to determine a Fundamental Reference Frame for each of the ten spaces (universes) of the HyperCosmos. We are motivated in this endeavor by the observation made in previous books (and later) that the fundamental fermion particles of a space (universe) appear to be expressible as sets of repeated fermions. We also see a similar duplication in the set of symmetries of a space. They are also expressible as sets of repeated products of symmetries. Sections 8.1 and 8.2 show this for an N = 7 space, which we have considered in our discussions of QUeST and the Unified SuperStandard Theory (UST). Remarkably, the set of duplicated fermions includes the Standard Model fermions with an additional fourth generation. And the set of duplicated symmetries includes SU(3)⊗U(1)⊗SU(2)⊗U(1)⊗SO⁺(1, 3)—those of the Standard Model. *Thus we show that the Fundamental Reference Frame is ultimately based on a Standard Model set of fermions and symmetries.*

Based on these observations we generalize the sets of creation/annihilation operators seen in chapters 4 – 6 to support General Relativistic transformations that map a core set of fermions and of internal symmetries in a Fundamental Reference Frame to replicated sets of multiples of fermions, and to replicated sets of internal symmetries in a conventional reference frame. The key to this generalization is the use of hypercomplex formulations of creation/annihilation operators. Effectively they each become a hypercomplex-valued operator.

8.1 Set of Fermions Expressed as Repeated Sets of Fermions in QUeST and UST

The fundamental fermion spectrum was derived in Blaha (2018e) (and earlier books) from symmetry groups that followed from (broken) fermion conservation laws. It consisted of 256 fermions arranged in four generations in four layers. of "Normal" fermions and an identical spectrum of Dark fermions. Remarkably this spectrum was surprisingly, and encouragingly, found in hypercomplex spaces studies by this author in 2019-2022. See the resulting book: Blaha (2020a). Since then the author has developed hypercomplex spaces theories that eventually resulted in the HyperCosmos (HyperComplex Cosmology) found recently.

The form of the 256 fermion spectrum for our universe (N = 7) continues to hold in the HyperCosmos. It may be viewed as 16 replicates of a set of 16 Normal and Dark fermions of the species: electron, neutrino, three up-type quarks, and three down-type quarks. Each of 16 fermion sets has its own set of quantum numbers and its own set of interactions. We are familiar with one level consisting of three generations of The

Standard Model, which is incorporated within the UST. The UST and HyperCosmos imply *four* generations.

The 16-fold replication of the 16 fermion sets leads us to consider the possibility that they are the result of a replication process. We found a formulation of the replication process based on General Relativity applied to creation/annihilation operators in a Fundamental Reference Frame discussed later. The formulation applies to all spaces of the HyperCosmos. All spaces of the HyperCosmos exhibit patterns of replication. The replication process necessarily applies to them as well.

	NORMAL		DARK	
	e q-up	**v q-down**	**e q-up**	**v q-down**
	e q-up	**v q-down**	**e q-up**	**v q-down**
Layer 1	**e q-up**	**v q-down**	**e q-up**	**v q-down**
	e q-up	**v q-down**	**e q-up**	**v q-down**
	e q-up	**v q-down**	**e q-up**	**v q-down**
	e q-up	**v q-down**	**e q-up**	**v q-down**
Layer 2	**e q-up**	**v q-down**	**e q-up**	**v q-down**
	e q-up	**v q-down**	**e q-up**	**v q-down**
	e q-up	**v q-down**	**e q-up**	**v q-down**
	e q-up	**v q-down**	**e q-up**	**v q-down**
Layer 3	**e q-up**	**v q-down**	**e q-up**	**v q-down**
	e q-up	**v q-down**	**e q-up**	**v q-down**
	e q-up	**v q-down**	**e q-up**	**v q-down**
	e q-up	**v q-down**	**e q-up**	**v q-down**
Layer 4	**e q-up**	**v q-down**	**e q-up**	**v q-down**
	e q-up	**v q-down**	**e q-up**	**v q-down**

Figure 8.1. Four layers of fermions with each layer containing four generations. The fermions are arranged in SU(4)-like sets of four fermions based on the structures specified in chapter 10. The sets could be arranged to a different grouping without changing the discussion in this chapter.

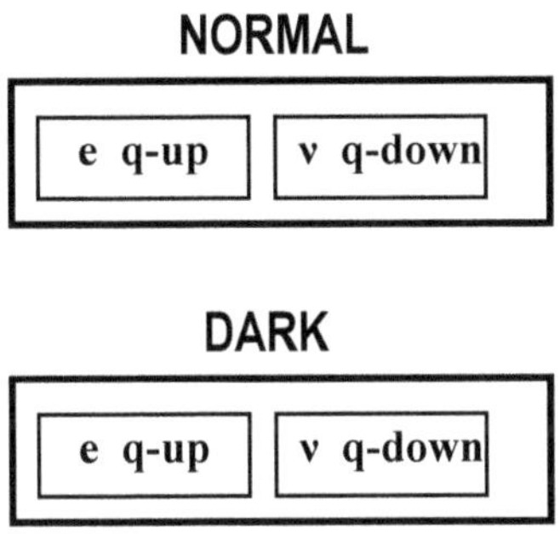

Figure 8.2. The 16 fermions that constitute the set that is replicated 16 times in the 256 fermion spectrum for N = 7 space (universe).

The separation of the 16 fermions into sets of four fermions in Fig. 8.2 and the form of Fig. 8.1 reflects the separation of fermions and internal symmetries into 4, 8, 16 32, and 64 parts due to symmetry splitting as we discussed in Blaha (2021b). We return to the discussion of symmetry splitting in chapter 11.

8.2 Set of Symmetry Groups Expressed as Repeated Sets of Groups in QUeST and UST

The set of symmetry groups in QUeST for our universe also exhibits the same sort of replication seen for fundamental fermions. The symmetry groups have a 256 real-valued dimensions representation as the product of the fundamental representations of their component groups. The basic symmetry group set for an N = 7 space is a U(4)⊗U(4) group. After a 16-fold replication the U(4) factors undergo transformations/breakdowns to[58]

U(4) or SU(3)⊗U(1) or SU(2)⊗U(1)⊗SL(2, **C**)[59] or SU(2)⊗U(1)⊗U(2)

Initially the N = 7 full set of groups has the form of 16 U(4)⊗U(4) symmetry groups (Fig. 8.3.) Then they are transformed by symmetry breaking to the groups in Fig. 8.4.

---

[58] In chaper 9 we raise the possibility that U(4)'s may undergo a breakdown via the Higgs Mechanism.
[59] We use SL(2, C) to represent $SO^{+}(1, 3)$.

	NORMAL	DARK
Layer 1	U(4)⊗U(4) U(4)⊗U(4)	U(4)⊗U(4) U(4)⊗U(4)
Layer 2	U(4)⊗U(4) U(4)⊗U(4)	U(4)⊗U(4) U(4)⊗U(4)
Layer 3	U(4)⊗U(4) U(4)⊗U(4)	U(4)⊗U(4) U(4)⊗U(4)
Layer 4	U(4)⊗U(4) U(4)⊗U(4)	U(4)⊗U(4) U(4)⊗U(4)

Figure 8.3. The "initial" distribution of sets of N = 7 symmetry groups. Each set is distinct and supports interactions only for the corresponding set of fermions (separately for Normal and Dark fermions) in Fig. 8.1. *Thus each set of 16 fermion generations has its own quantum numbers and interactions.* Each U(4)⊗U(4) set has a 16 real-valued dimension representation, which will be of importance when we consider Fundamental Reference Frames in section 8.3.

NORMAL		DARK	
SU(3)⊗U(1) Generation U(4)	SU(2)⊗U(1)⊗SL(2, C) Layer U(4)	SU(3)⊗U(1) Generation U(4)	SU(2)⊗U(1)⊗U(2) Layer U(4)
SU(3)⊗U(1) Generation U(4)	SU(2)⊗U(1)⊗U(2) Layer U(4)	SU(3)⊗U(1) Generation U(4)	SU(2)⊗U(1)⊗U(2) Layer U(4)
SU(3)⊗U(1) Generation U(4)	SU(2)⊗U(1)⊗U(2) Layer U(4)	SU(3)⊗U(1) Generation U(4)	SU(2)⊗U(1)⊗U(2) Layer U(4)
SU(3)⊗U(1) Generation U(4)	SU(2)⊗U(1)⊗U(2) Layer U(4)	SU(3)⊗U(1) Generation U(4)	SU(2)⊗U(1)⊗U(2) Layer U(4)

Figure 8.4. The transformed/broken sets of symmetries in QUeST (UST) and in N = 7 HyperCosmos space.. Note each element has a 16 real dimension representation. This depiction is also evident in QUeST and the UST. The SL(2, **C**) representation has four coordinates.[60] Seven of these groups are transformed to U(2) Connection groups.

8.2.1 Are there Dynamics in the Fundamental Reference Frame?

Since the Fundamental Reference Frame has symmetry groups of the form U(4)⊗U(4) there is no time in it[61]—no dynamical evolution. After the transformation[62] to a static reference frame, an SL(2, C) is generated with time evolution as a result.

[60] The Lorentz Group $SO^{+}(1, 3)$ is often specified with an SL(2, **C**) representation.

[61] It is a non-static frame with no time defined.

[62] The transformation from the Fundamental Reference Frame to a static frame might be designed to embody symmetry breaking with Fig. 9.4 the result.

8.3 General Relativistic Transformation from a Fundamental Reference Frame to a Static Coordinate System

We now develop a formulation for the transformation of a core set of fermions and symmetry groups to sets of duplicates such as those described in sections 8.1 and 8.2. The formulation will apply to all HyperCosmos spaces (universes). This formulation was first presented in Blaha (2021j).

We begin by generalizing the sets of creation/annihilation operators to support General Relativistic transformations that map a core set of fermions and of internal symmetries in a Fundamental Reference Frame to replicated sets of fundamental fermions and of replicated sets of internal symmetries residing in a conventional static reference frame. The key to this generalization is the use of hypercomplex extensions of creation/annihilation operators. Each normal creation/annihilation operator effectively becomes a hypercomplex operator.

We will consider the general case of the spaces and universes of Blaha number N. (The particular case of Blaha number 7 (our space and universe) was considered in some detail above and in Blaha (2021j).

The number of elements in the dimension array d_{dN} for Blaha number N is[63]

$$d_{dN} = 2^{22-2N} = 2^{r+4} \tag{4.67}$$

where r is the space-time dimension. It satisfies

$$N = Os = ½(18 - r) \tag{2.30}$$

The dimension array is a square array whose rows and columns have 2^{11-N} elements. The associated CASe group is su($2^{r/2}$, $2^{r/2}$). Its representation, which is composed of the direct product of fundamental representations, has a real-valued vector w_N with 2^{11-N} elements. We can also define vector v_N with the $d_{dN} = 2^{22-2N}$ elements of the dimension array. Note that the elements of v_N can be placed in a one-to-one correspondence with the b's and d's of the fermion with the least half-integer spin in the space of Blaha number N. The vector v_N can be placed in a one-to-one correspondence with the set of fundamental fermions of the space, and with the set of dimensions of the fundamental representations of the symmetries of the space (including space-time dimensions.)

Now we note that General Relativistic transformations may take an individual b or d creation/annihilation operator and transform it using a CASe group su($2^{r/2}$, $2^{r/2}$) transformation with a 2^{11-N} by 2^{11-N} rows/columns matrix representation to a sum of creation/annihilation operators for each momentum.

We take this concept and use it to relate the vectors w_N to the vectors v_N. To do so, we must *initially* use a reducible representation for su($2^{r/2}$, $2^{r/2}$) with a matrix representation with a $2^{22-2N} = 2^{r+4}$ by $2^{22-2N} = 2^{r+4}$ rows/columns. The vector v_N has 2^{22-2N} components. However w_N has only 2^{11-N} components.

---

[63] Fig. 1.1.

We therefore define an enhanced vector W_N with 2^{22-2N} components by mapping each element of w_N from a real-valued number to a hypercomplex number with 2^{11-N} elements.[64] Each hypercomplex number may be viewed as composed of one non-zero element followed by $2^{11-N} - 1$ elements containing zeroes. The transformation of eq. 8.1 creates a new vector v_N with all components non-zero in general. This vector is a static space-time coordinate vector for fermions or symmetries or the dimensions of the dimension array d_{dN}.

We augment the General Relativistic transformation by generalizing su($2^{r/2}$, $2^{r/2}$) to su(2^r, 2^r). It has a matrix representation with $2^{22-2N} = 2^{r+4}$ by $2^{22-2N} = 2^{r+4}$ rows and columns.[65] Thus we have the transformation

$$v_N = [T]\, W_N \tag{8.1}$$

where T has a $2^{22-2N} = 2^{r+4}$ by $2^{22-2N} = 2^{r+4}$ matrix form. *The form of [T] is equivalent to $2^{22-2N} \times 2^{11-N}$ array which maps a 2^{11-N} vector to a 2^{22-2N} vector.*

For example in a universe of an N = 7 space the 16 elements in the FRF are mapped by a 256×16 transformation array to a 256 element vector that is separated into a 16 × 16 square array format for fermions, and for the real parts of the dimensions of irreducible symmetry group representations. See chapter 9 for details.

We may take the vector space of the w_N to be a non-static Fundamental Reference Frame since almost all reference frames are non-static.

The group su(2^r, 2^r) is beyond the General Relativistic-based group su($2^{r/2}$, $2^{r/2}$). By augmenting the reducible su($2^{r/2}$, $2^{r/2}$) representation to make an irreducible su(2^r,2^r) representation, we require transformations from the Fundamental Reference Frame to a static frame to include General Relativity *and internal symmetry group aspects. The su(2^r,2^r) group unites General Relativistic transformations with internal symmetry transformations.*

8.4 Contents of the Fundamental Reference Frame

The contents of a Fundamental Reference Frame (FRF) number 2^{11-N} fermions or dimensions or symmetry dimensions (for each symmetry's fundamental representation.) In each case these items are distributed within the hypercomplex coordinates introduced into the frame. For example, for N = 7 there are 16 fermions residing within 16 hypercomplex (sedenion) coordinates with the vector W_7 thus containing 256 real-valued coordinates.

The number of physical elements in the Fundamental Reference Frame of space N is 2^{11-N} and the total number of elements in the hypercomplex vector containing them is 2^{22-2N}. A transformation of the type of eq. 8.1 causes the generation of a 2^{22-2N} element vector representing the replication of fermions or internal symmetry

[64] In chapter 10 we introduce a higher dimension unification space which implements the equivalent of this procedure of replacing raising and lowering operators with hypercomplex forms. For example, the unification space for our universe has a 12 dimension space-time.

[65] The transformation now is a General Relativistic – Internal Symmetry transformation.

dimensions or dimensions in general. Thus the transformation effectively "squares" the number of physically relevant quantities.

One can view a Fundamental Reference Frame as the "rest" frame for a HyperCosmos space. A generalized General Relativistic transformation maps the "rest" frame to a static coordinate system with replicates of the contents of the rest frame.

Fig. 8.5 below tabulates the contents of the ten Fundamental Reference Frames of the ten HyperCosmos spaces (universes).

Contents of Fundamental Reference Frames

N	Number of Physical Elements FRF	Total Number of Elements Generated
0	2^{11}	2^{22}
1	2^{10}	2^{20}
2	2^{9}	2^{18}
3	2^{8}	2^{16}
4	2^{7}	2^{14}
5	2^{6}	2^{12}
6	2^{5}	2^{10}
7	2^{4}	2^{8}
8	2^{3}	2^{6}
9	2^{2}	2^{4}

Figure 8.5. The contents of the ten types of Fundamental Reference Frames for the ten spaces. The elements are fermions or symmetry fundamental representation dimensions or dimensions.

The contents of a Fundamental Reference Frame are fermions or symmetries or dimensions. We display the contents for the case of fermions for N > 5 in Fig. 8.6. Note the N = 9 space has units of fermions composed of four fermions. The number of units repeatedly doubles as we ascend up to N = 0.

8.5 Contents of the Higher Dimensions FRF Frame

The FRF reference frame generated by the transformation in eq. 8.1 has 2^{11-N} copies of the 2^{11-N} items in the Fundamental Reference Frame. For example for N = 7 (our universe) 16 copies are made of the 16 items in its Fundamental Reference Frame as seen earlier.

Thus the 2^{22-2N} elements of the dimension array (or the fermion array obtained from it) consist of 2^{11-N} copies of the 2^{11-N} physical elements in the Fundamental Reference Frame.

The 2^{11-N} physical elements consist of 2^{7-N} replicates of the N = 7 set of 16 physical elements for N < 7

Contents of a Fundamental Reference Frame (FRF) for Fermions

N	Number of Physical Elements	Fermion Content	
0	2^{11}		
1	2^{10}		
2	2^{9}		
3	2^{8}	•	
4	2^{7}	•	
5	2^{6}	•	
6	2^{5}	NORMAL e q-up v q-down e′ q-up′ v′ q-down′ DARK e q-up v q-down e″ q-up″ v″ q-down″	Total: 32 fermions
7	2^{4}	Normal: e q-up v q-down Dark: e q-up v q-down	16 fermions
8	2^{3}	e q-up AND v q-down	8 fermions
9	2^{2}	e q-up OR v q-down	4 fermions

Figure 8.6. Possible contents of five of the ten types of Fundamental Reference Frames for fermions. The primes distinguish different fermions.

The symmetry group contents in the Fundamental Reference Frame are listed for N > 5 in Fig. 8.7. Note the units of symmetry irreducible representations dimensions are four real-valued dimensions. Irreducible symmetry group units also have four dimensions. The number of units repeatedly doubles as we ascend up to N = 0.

Contents of a Fundamental Reference Frame for Symmetry Group Dimensions

N	Number of Physical Elements	Fermion Content		
0	2^{11}			
1	2^{10}			
2	2^{9}			
3	2^{8}	•		
4	2^{7}	•		
5	2^{6}	•		
6	2^{5}	NORMAL U(4)⊗U(4) DARK U(4)⊗U(4)		Total: 32 real dimensions
7	2^{4}	NORMAL U(4)	DARK U(4)	16 real dimensions
8	2^{3}	U(4)		8 real dimensions
9	2^{2}	U(2)		4 real dimensions

Figure 8.7. Possible contents of five of the ten types of Fundamental Reference Frames for symmetry dimensions. The groups may undergo a further breakdown after being mapped to a static space-time to give terms like those in Fig. 8.4.

We describe the generation of the form of FRF symmetries in chapter 9.

8.6 Meaning of Space-Time in the HyperCosmos

Each space-time within the HyperCosmos spectrum of spaces plays an important role. The question arises: What sets the number of dimensions in a space-time? In this section we answer that question and show a direct connection to the Fundamental Reference Frames. We begin by noting

$$r = 18 - 2N \qquad (8.2)$$
$$d_{dN} = 2^{22-2N} = 2^{r+4}$$

where r is the dimension of the space-time, and d_{dN} is the number of elements in the dimension array for Blaha space N. These equations imply

$$r = \log_2 (d_{dN}/16) \qquad (8.3)$$

Noting that d_{dN} is the square of the number of physical elements d_{FRF} in the Fundamental Reference Frame.:

$$d_{dN} = d_{FRF}^{2} \tag{8.4}$$

we see

$$r = 2 \log_2 (d_{FRF}/4) \tag{8.5}$$

or, in words, r is twice the logarithm (base 2) of the total number of units of four elements of physical fermions or symmetries or dimensions in the non-static Fundamental Reference Frame. Note the dimensions in the dimension array d_{dN} include the dimensions of all the internal symmetry irreducible representations and the space-time dimensions.

Thus the space-time dimension of each space may be viewed as determined ultimately by its Fundamental Reference Frame physical elements. The number of space-time dimensions in a space, the total number of internal symmetry dimensions of a space, and the number of fundamental fermions of the space's Fundamental Frame are linked thus supporting the reality of the Fundamental Reference Frame.

This chapter has shown how the sets of fermions and dimensions are generated by a CASe transformation due to a General Relativistic – Internal Symmetry transformation of the fermions and dimensions from the Fundamental Reference Frame to a static reference frame

8.7 A Physical Implementation of Platonic Realism: Plato's Theory of Abstract Forms

Platonic Realism views the universe as based on universals or abstract objects, which are called Ideal Forms. The world (universe) is described as imperfect dynamic embodiments of these unchanging abstract forms. The Fundamental Reference Frames that we have found may be viewed as implementing the ideal forms of Plato. A Fundamental Reference Frame has fermions, dimensions and symmetries defined as abstract entities. They are fixed, and unchanging, since time and interactions are not executing dynamically. Fundamental Reference Frames are non-static reference frames with no defined time evolution.

HyperCosmos spaces are defined for static reference frames. They are the dynamically changing, imperfect embodiments of the Fundamental Reference frames "Ideal Forms."

Each of the ten spaces of the Hyper Cosmos has a set of Fundamental Reference Frames. The contents of the frames show a progression in the sense that the contents of the FRF of the space of Blaha number N is twice the contents of the FRF of space N + 1.

The requirements for a physical analogue of Platonic theory of Realism therefore appear to be met in a physical context by Hyper Cosmos spaces and their Fundamental Reference Frames.

9. Space-Time Determination of Particles and Symmetries

We now turn to consider the form of fermion species due to the appearance of two times in su(1, 1).[66] The su(1, 1) CASe group determines the form of the species of fermions in all HyperCosmos spaces since it is present in all spaces.

The su(1, 1) group has a metric with two real time coordinates:

$$ds^2 = t_{01}{}^2 + t_{02}{}^2 - x_1{}^2 - x_2{}^2 \tag{9.1}$$

In Blaha (2007b) we showed that the four species of fermions: e-type, ν-type, q-up-type and q-down-type followed from the four types of boosts of the Lorentz Group $SO^+(1, 3)$, which is often expressed as a representation of SL(2, **C**)). Complex Lorentz group boosts separate into sublight, superluminal, complex sublight, and complex superluminal boosts yielding the four fermion species respectively.

Now we have a more interesting situation with two time coordinates.[67] Each has its own light speed. We take them to be numerically equal for the purposes of our discussion. One light speed divides the set of possible fermions into two "superspecies", namely Normal and Dark fermions. The second light speed further divides each superspecies into four parts. The result is the eight fermion species *in our universe, Fig. 9.1.* It includes both Normal and Dark sectors. It is also the form of fermion species in the other HyperCosmos spaces which also have su(1, 1) multiple time coordinates.

The 16 fermions of an N = 7 Fundamental Reference Frame can be separated into the eight subspecies if account is taken of the occurrence of each quark species as a triplet. This separation of fermion species leads to the structure of the Unified SuperStandard Theory (UST) and the Standard Model. It also leads to the form of Fundamental Reference Frames for all HyperCosmos spaces as described in chapter 8.

A complete, detailed spectrum of fundamental fermions in our universe, and other universes, thus emerges, part of which appears in the Standard Model for our universe.

9.1 The Four Fermion Species Becomes Eight Subspecies for N = 7

In Blaha (2007b) we showed that four species of fermions: e-type, ν-type, q-up-type and q-down-type followed from the four types of boosts of the Lorentz Group $SO^+(1, 3)$, which is often expressed as SL(2, **C**)). Complex Lorentz group boosts separate into sublight, superluminal, complex sublight, and complex superluminal boosts yielding the four species respectively for both Normal and Dark sectors. Now we

[66] See chapter 4.

[67] Chapter 9 describes the transformation between a Fundamental Reference Frame where superluminal quantities are present and a static frame where they are transformed to sublight quantities.

extend the form to two time coordinates in Fig. 9.1. Each of the 16 FRF dimensions is mapped to one fermion in the FRF giving 16 FRF fermions. Subsequently 16 replicas of each FRF fermion are generated by a General Relativistic – Internal Symmetry transformation as described in chapter 8. The sixteen replicas assemble into four layers, with each consisting of four generations of fermions.

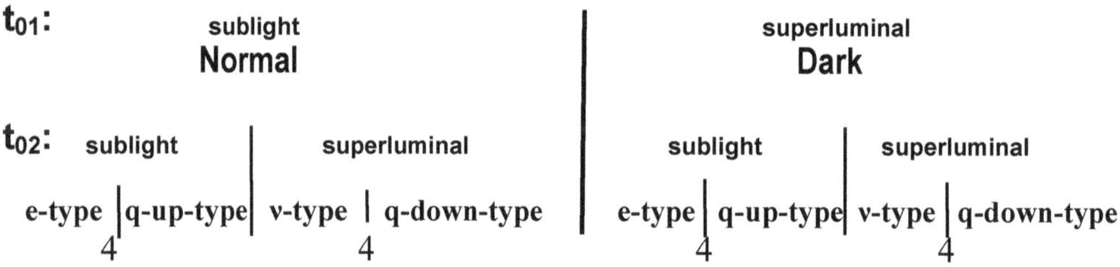

Figure 9.1. Separation of fermion species due to su(1, 1) and its two times. The total number of corresponding fermions is 16 if account is taken of quarks appearing as triplets. The predicted occurrence of quark in triplets in the HyperCosmos is motivated by the need for fermions to appear in multiples of four units. Thus the separation by two light speeds may account for the Standard Model form of the fermion species and also for triplets of quarks. Experimentally, superluminal neutrinos appear to move at the speed of light due to their negligible masses. Down-type quarks, being confined within particles, do not directly display their superluminal nature. Dark matter, being unobserved may well be superluminal. Their superluminal nature may be part of the cause for their current unobservability. It is also possible that the superluminal nature of these quantities may "wash out" when a transformation is made from an FRF to a static reference frame.

9.2 Scalar Bosons of HyperCosmos Symmetries for N = 7

We now consider scalar bosons including Higgs bosons. Unlike the fermion case the irreducible representations of the symmetry groups have complex dimensions that may be separated into real and imaginary parts. The set of 8 scalar boson complex irreducible representation dimensions are separated into 16 parts. The real parts (or imaginary parts) can be used to form N = 7 FRF sets of superluminal (Dark) and subluminal (Normal) sets of 8 based on the two times coordinates. (See Fig. 9.2a.)

A General Relativistic – Internal Symmetry transformation of the type[68] described in chapter 8 results in 16 Normal and 16 Dark scalar boson complex representation dimensions, which support scalar boson wave function generation in each of four layers where each boson is a combination of a real and imaginary part.

In each Normal layer there is a non-Higgs set of three SU(4) representation bosons and two sets of Higgs bosons with each containing two bosons in an SU(2) representation. The Dark sector has similar sets of bosons. See Fig. 9.2b.

[68] The transformation is a 16 × 16 array.

t_{01}:	sublight **Normal** U(4)		superluminal **Dark** U(4)	
real wave function parts	8		8	
t_{02}:	sublight	superluminal	sublight	superluminal
	Non-Higgs	Higgs Bosons	Non-Higgs	Higgs Bosons
Parts	6 SU(3)	2 SU(2)	6 SU(3)	2 SU(2)

Figure 9.2a. Real parts of bosons separated by su(1, 1)'s two times. Each set of two parts will combine, as real and imaginary parts, into a scalar boson in a General Relativistic – Internal Symmetry transformation.

In the ***Normal*** sector the transformation's 16-fold replication gives[69]

$8 \times 16 = 128$ real parts combine to create
64 bosons giving
16 bosons per layer with the result
SU(4)⊗SU(4)⊗SU(4)⊗SU(2)⊗SU(2) bosons in each layer.

with a similar result for the Dark sector. The result is 64 Normal scalar bosons and 64 Dark scalar bosons which total to 128 scalar bosons/

These sets of scalar bosons are assigned to each of the four layers as seen earlier for the fundamental fermions. Thus each layer has five sets of scalar bosons in each of the Normal and Dark sectors: two sets of SU(2) Higgs scalar bosons and three sets of non-Higgs SU(4) scalar bosons. The "superluminal" Higgs bosons each have a negative mass squared.

The three sets of subluminal SU(4) bosons in the Normal and Dark sectors of each layer do not undergo the Higgs mechanism since these "subluminal" bosons each have a positive mass squared.

Higgs bosons have the remarkable feature that they are tachyonic (superluminal) due to their negative mass squared terms. Their tachyonic nature is masked by quartic terms that result in Higgs Mechanism symmetry breaking.

The distinction between tachyonic and non-tachyonic (subluminal) bosons and symmetries seen below is addressed again in chapter 12 where we suggest that Gravitation and the Strong interactions have higher derivative interactions because they are subliminal while ElectroWeak-type interactions do not have higher derivative

[69] The transformation for Normal and Dark sectors corresponds to a 16×16 array.

interactions because they are superluminal *due to the structuring of mass terms* in the full Lagrangian.[70] We suggested this distinction some years ago.

Scalar bosons are replicated similarly to the fermions to produce a total of 128 Normal and Dark scalar bosons (of differing symmetries) by a General Relativistic – Internal Symmetry transformation.

t_{01}:	sublight **Normal** **U(4)**	superluminal **Dark** **U(4)**
Real-valued wave function parts	**8**	**8**

t_{02}:	sublight **Non-Higgs**	superluminal **Higgs Bosons**	sublight **Non-Higgs**	superluminal **Higgs Bosons**
	6	**2**	**6**	**2**
Complex	SU(3)	SU(2)	SU(3)	SU(2)

Transformation

Level					
1	Non-Higgs	Higgs	Non-Higgs	Higgs	
	SU(4)	SU(2)	SU(4)	SU(2)	(I)
	SU(4)	SU(2)	SU(4)	SU(2)	(II)
	SU(4)		SU(4)		(III)
2	Non-Higgs	Higgs	Non-Higgs	Higgs	
	SU(4)	SU(2)	SU(4)	SU(2)	(I)
	SU(4)	SU(2)	SU(4)	SU(2)	(II)
	SU(4)		SU(4)		(III)
3	Non-Higgs	Higgs	Non-Higgs	Higgs	
	SU(4)	SU(2)	SU(4)	SU(2)	(I)
	SU(4)	SU(2)	SU(4)	SU(2)	(II)
	SU(4)		SU(4)		(III)
4	Non-Higgs	Higgs	Non-Higgs	Higgs	
	SU(4)	SU(2)	SU(4)	SU(2)	(I)
	SU(4)	SU(2)	SU(4)	SU(2)	(II)
	SU(4)		SU(4)		(III)

Figure 9.2b. The sets of scalar bosons generated by the transformation in the four layers. Line (I) is for the Standard Model sector enhanced to four generations and a Dark sector. Line (II) corresponds to the Generation group

[70] We presented this idea some time ago. See Blaha (2018e) for a discussion in the UST.

and an U(2) Connection Group described earlier. Line (III) corresponds to the Layer Group also described earlier. .

9.3 Symmetries Separation Due to Space-Time

The symmetries for the Fundamental Reference Frame also have representations comprised of irreducible group representations with totals equal to multiple units of 4 real dimensions. Again the su(1, 1) CASe group with its two time dimensions structures the set of groups. Fig. 9.3 displays the separation of the representations in sublight vs. superluminal parts.

9.4 N = 7 Fundamental Reference Frames Contents

In chapter 8 we constructed Fundamental Reference Frames. Their contents for fermions and symmetries are structured according to the formatting presented in this chapter. Thus we have a formulation for the general structure of the Fundamental Reference Frame for N = 7 universes, and also for N < 7 universes by extension.

The FRF contains the real parts of the complex dimensions of the irreducible representations of symmetry groups in a manner similar to the scalar boson case studied above.

9.5 Total View of N = 7 HyperCosmos Symmetries

The reduction of *each* HyperCosmos space to a Fundamental Reference Frame set of symmetries and fermions together with the partitioning of fermions and symmetries seen above in this chapter suggests a ***complete HyperCosmos structuring of the fermion spectrum and of the set of symmetries independent of the details of symmetry breaking and coupling constant values.***

t_{01}:	sublight NORMAL U(4) 8		superluminal DARK U(4) 8	
t_{02}:	sublight	superluminal	sublight	superluminal
	SU(3)	SU(2)	SU(3)	SU(2)
Real Dimensions	6	2	6	2

Figure 9.3. The real parts of dimensions in the N = 7 FRF's symmetry part. The separation of symmetries is due to su(1, 1) and its two times. The separation by two light speeds accounts for the UST (and Standard Model) form of the interaction groups.

After a General Relativistic – Internal Symmetry transformation[71] the set of QUeST/UST internal symmetries results. See Figs. 1.5, 7.4, 7.5 8.4 and 8.5.

[71] The transformation for Normal and Dark sectors corresponds to a 16 × 16 array.

The ***Normal*** symmetries sector of QUeST/UST generated from the 16 FRF symmetry components of Fig. 9.3 is based on the calculation:

$8 \times 16 = 128$ real dimensions combine to create
64 complex dimensions giving
16 complex dimensions per layer
$SU(4)\otimes U(4)\otimes U(4)\otimes SU(2)\otimes U(1)\otimes U(2)$ bosons in each layer.[72]

The calculation parallels the scalar boson calculation above. The Generation and Layer groups are U(4) groups because each has four diagonal particle numbers.[73] Fig. 9.4 displays the symmetries generated by the transformation.

9.6 Extension to Higher Dimension Spaces and FRF's

Section 8.5 showed the 2^{11-N} physical elements of the space N FRF consists of 2^{7-N} replicates of the space N = 7 set of 16 physical elements. Each of these FRF replicates has their own symmetries.

On this basis we see the above considerations of the two times structuring for N = 7 extend directly to each of the 2^{7-N} replicates in the FRF of space N.

A General Relativistic – Internal Symmetry transformation [T] maps FRF elements to a reference frame containing the full set of elements. The elements, as seen above, form a 2^{11-N} vector in the FRF. They are transformed to a 2^{22-2N} vector in a physical coordinate system. This vector is then changed into a $2^{11-N} \times 2^{11-N}$ square array of fermions, or the real parts of scalar, bosons irreducible representations, or the real parts of symmetry groups' irreducible representations as seen above.

The form of [T] is equivalent to $2^{22-2N} \times 2^{11-N}$ *array which maps a* 2^{11-N} *vector to a* 2^{22-2N} *vector. The* 2^{11-N} *vector is transformed to a* 2^{22-2N} *vector and [T] is transformed to* $2^{22-2N} \times 2^{22-2N}$ *array to implement the transformation in this chapter. Chapter 10 simplifies the process using a unification space.*

---

[72] Layer 1 replaces a U(2) with SL(2, **C**).
[73] See Appendix 7-A.

t_{01}:	sublight **NORMAL** **U(4)** **8**		superluminal **DARK** **U(4)** **8**	
t_{02}:	sublight	superluminal	sublight	superluminal
	SU(3)	**SU(2)**	**SU(3)**	**SU(2)**
Real Dimensions	**6**	**2**	**6**	**2**

Transformation

Level

Level					
1					
	SU(4)	SU(2)⊗U(1)	SU(4)	SU(2)⊗U(1)	(I)
	U(4)	SL(2,**C**)[74]	U(4)	U(2)	(II)
	U(4)		U(4)		(III)
2					
	SU(4)	SU(2)⊗U(1)	SU(4)	SU(2) ⊗U(1)	(I)
	U(4)	U(2)	U(4)	U(2)	(II)
	U(4)		U(4)		(III)
3					
	SU(4)	SU(2)⊗U(1)	SU(4)	SU(2) ⊗U(1)	(I)
	U(4)	U(2)	U(4)	U(2)	(II)
	U(4)		U(4)		(III)
4					
	SU(4)	SU(2)⊗U(1)	SU(4)	SU(2) ⊗U(1)	(I)
	U(4)	U(2)	U(4)	U(2)	(II)
	U(4)		U(4)		(III)

Figure 9.4. The contents of the N = 7 symmetries. The separation of symmetries due to su(1, 1) with two times. The separation by two light speeds accounts for the QUeST/UST (and the Standard Model) form of the interaction groups. Line (I) is for the Standard Model sector enhanced to four generations and to a Dark sector. Line. (II) corresponds to the Generation group, and an U(2) Connection Group[75] described earlier except line (II) of layer 1 which has a space-time group SL(2, **C**). Line (III) corresponds to the Layer Group also described earlier.

[74] $SO^{+}(1, 3)$.

[75] See Appendix 7-B.

10. Unification: HyperCosmos and F-Theory

This chapter demonstrates[76] the unification of General Relativity and Internal Symmetries in all universes of the ten HyperCosmos spaces. Each space is mapped to a unification space of the same form as the HyperCosmos spaces. In the case of our N = 7 space the unification space has 12 dimensions and two time coordinates. It is a particular case of F-Theory with the special feature that the sets of fundamental fermions and symmetries are fully determined. (See chapters 8 and 9.). The contents of other spaces and their unification spaces are also fully determined.

10.1 The HyperCosmos Space and Unification Space of Our Universe

The HyperCosmos, that we discovered,[77] has a number of noteworthy features for the case of our universe (Blaha number N = 7):

1. 4 space-time dimensions.

2. A 12 dimension unification space-time of the form of a HyperCosmos space..

3. Two time coordinates in the 12 dimension unification space-time[78] that separate the fermions into e-type, ν-type, up-quark-type, and down-quark-type species, and also separate internal Standard Model symmetries into SU(3), U(1) and SU(2)⊗U(1) groups.

4. The HyperCosmos implies QUeST for our universe. QUeST has 256 gauge vector field real-valued irreducible representation dimensions and 256 fundamental fermions.

The net effect of these features leads us to view the 12 space-time dimension space as an F-Theory, which leads to a type IIB SuperString theory.

Given the above points the question arises: What larger F-Theory analogue (with its corresponding SuperString theory) corresponds to the N = 6 HyperCosmos space (our Megaverse) with 6 space-time dimensions and a 16 space-time dimension unification space?

In addition, the other seven HyperCosmos spaces also raise the same question of F-Theory analogues.[79] These questions have not been unexplored.

---

[76] This chapter appeared in Blaha (2022e) in a somewhat modified form.
[77] Blaha (2022c) and earlier books.
[78] See Chapter 10 in Blaha (2022c).
[79] See chapter 5 0f Blaha (2022d) for a derivation of the unification of HyperCosmos spaces.

10.2 HyperCosmos Grand Unification

The HyperCosmos *self*-generates its sets of internal symmetries. Each HyperCosmos space has a dimension array that specifies the complete set of internal symmetries and space-time, specifies the complete set of fundamental fermions, specifies the complete set of gauge vector bosons, and the complete set of Higgs bosons. Ea*ch space specifies a Grand Unified Theory (GUT)* and has a *Unification* subsidiary space although the possibility of additional fermions and interactions is not excluded.

Each HyperCosmos space has a space-time dimension r that is related to the Blaha number N of the space:

$$N = ½(18 – r) \qquad (10.1)$$

Our universe has r = 4 and N = 7.

The dimension array d_{dN} of a space N has a total of

$$d_{dN} = 2^{22 -2N} = 2^{r+4} \qquad (10.2)$$

elements in its dimension array. In one map each element maps to a fundamental fermion. In another map each element becomes a real-valued fundamental representation dimension. For example, SU(3) has 6 real-valued fundamental representation dimensions.

The dimension array of a space mapped to group representation dimensions contains both the space-time dimensions of the space (four for our universe) and the internal symmetries of the space. For our universe with d_{dN} = 256: there are 4 space-time dimensions and 252 internal symmetry fundamental representation dimensions.

In Blaha (2022e) and (2022e) we presented a Unification for our universe based on a map to a higher space-time dimension r' HyperCosmos space.[80] The higher space has a General Relativity that generates transformations that map to forms of the dimension array $d_{dN'}$ where N' is the Blaha number of the higher space. The higher space and the "lower" space have space-time dimensions related by:

$$r' = 2r + 4 \qquad (10.3)$$

based on the relation of the r dimension transformation and the r' dimension transformation. To obtain the dimension array for the r coordinate space-time we can use the r' coordinate space-time. The r' space-time transformation is a $2^{r'/2+2} \times 2^{r'/2+2} = 2^{r+4} \times 2^{r+4}$ array that generates a $2^{r'/2+2} = 2^{r+4}$ coordinates vector. This vector maps to the r space-time dimension array d_{dN}.[81] The transformations are General Relativistic

[80] See Blaha (2022d) and02022e) for more detail. The higher dimension space may be one of the ten HyperCosmos spaces or it may be another HyperCosmos-like space. The unification spaces have space-time dimensions ranging up to 40 dimensions. See Fig. 10.1.

[81] The array is divided into a square d_{dN} array.

transformations of the higher space. *These transformations embody both the General Relativistic space-time and Internal Symmetries of the lower space.*

The $2^{r'/2+2}$-vector elements of the higher space become the elements of the d_{dN} dimension array of the "lower" space. For example, in our four dimension universe the higher space has 12 space-time coordinates. Its dimension array has 256×256 = 65,536 elements.[82] It has a 256-vector that becomes the d_{dN} = 256 elements dimension array[83] of our universe with 4 space-time dimensions and 252 internal symmetry dimensions. The higher space-time dimension procedure yields the r space-time dimension array. *The key role of the higher space is Unification in this discussion.*

Note:

We have used the space corresponding to r' space-time dimensions to create the r space-time dimension array. The dimension array in each space-time specifies the number of fermion creation and annihilation operators in that space-time.

The number of elements in the higher space of space-time dimension r' dimension array is:

$$d_{dN}' = 2^{r'+4} \tag{10.4}$$

THE HYPERCOSMOS SPACES SPECTRUM

Blaha Space Number $N = O_s$	Cayley-Dickson Number n	Cayley Number d_c	Dimension Array column length d_{cd}	Dimension Array Size d_{dN}	Space-time-Dimension r	Higher Space-time r' Source of d_{dN} r'	Higher Space Array d_{dN}'
0	10	1024	2048	2048^2	18	40	2^{44}
1	9	512	1024	1024^2	16	36	2^{40}
2	8	256	512	512^2	14	32	2^{36}
3	7	128	256	256^2	12	28	2^{32}
4	6	64	128	128^2	10	24	2^{28}
5	5	32	64	64^2	8	20	2^{24}
6	4	16	32	32^2	6	16	2^{20}
7	**3**	**8**	**16**	$\mathbf{16^2}$	**4**	**12**	$\mathbf{2^{16}}$
8	2	4	8	8^2	2	8	2^{12}
9	1	2	4	4^2	0	4	2^{8}

Figure 10.1. The HyperCosmos spaces spectrum related to a unification higher space space-time dimension and its dimension array. Note the spaces with r' > 18 are outside the HyperCosmos set of 10 spaces. However they have the same form as the 10 HyperCosmos spaces.

In terms of r' we find its Blaha number N' is

---

[82] The value of the elements is not relevant for our discussions. The fact that they are non-zero in general is the key point.

[83] Again we note the value of the elements is not relevant. They play a symbolic role in being mapped to fermions and mapped to symmetry group dimensions, and so on.

$$N^{\prime} = \frac{1}{2}(18 - r^{\prime}) = 7 - r \tag{10.5}$$

Note: negative values of $N^{\prime}$ are allowed for spaces where $r > 7$.

Thus the introduction of a higher dimension space, as in Fig. 10.2 for the HyperCosmos spaces, allows us to ***unify*** *the space-times (and consequently General Relativity) with the internal symmetries of a lower HyperCosmos space within the General Relativity of the higher dimension space-time.*

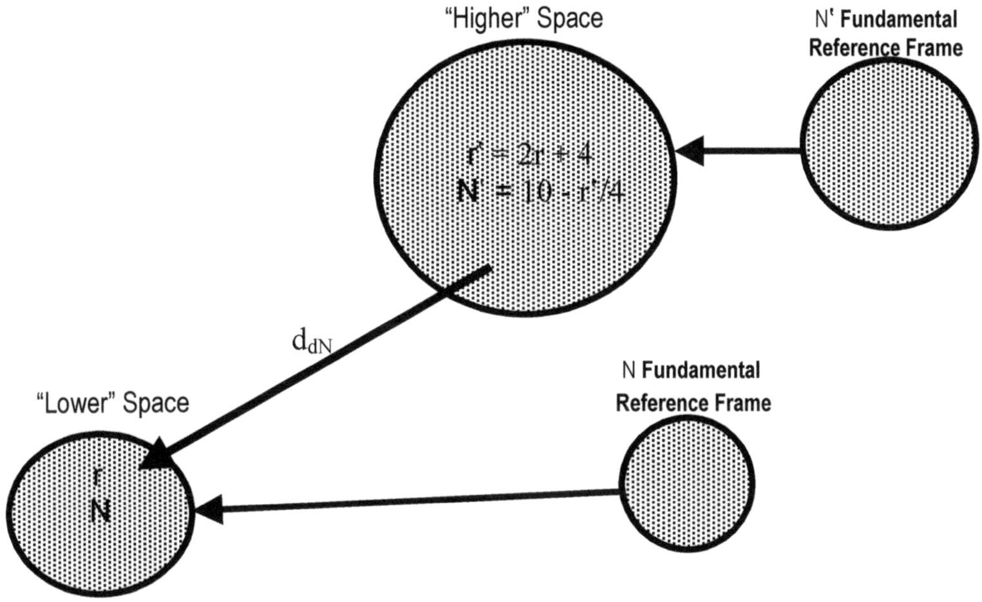

Figure 10.2. Diagram of the relation between a space and its "companion" higher space-time dimension space with Fundamental Reference Frames indicated. The higher space sets d_{dN} for the lower space using 'Higher' space General Relativity. Fundamental Reference Frames are like the rest frames of particles. They reduce the number of elements to a core set of elements. The elements are fermions or gauge vector symmetry fundamental representation dimensions.

The procedure for determining the lower space d_{dN} is outlined in Fig. 10.3.

1. Determine Higher Unification Space "partner" from the selected space
2. Find Higher Space FRF
3. Populate Higher Space FRF coordinates with a set of "Original" non-zero Coordinates
4. Generate the dimension array vector of the Selected Space with a General Relativistic Transformation
5. Partition the (dimension array) vector making a square array whose components number d_{dN}.

Figure 10.3 The steps for determining a dimension array using a General Relativistic transformation in the "partner" higher space. Note the numerical values generated by the higher space General Relativistic transfomation are only important in so far as they are zero or non-zero.

10.2.1 Form of Our Universe Dimension Array and Symmetries

The sets of fermions and internal symmetries follow from a pattern embodied in the hypercomplex aspects of the spaces. Items are divided into sets of 4, sets of 8, sets of 16, sets of 32 and sets of 64. This partitioning of dimension arrays (particles and symmetries) is described in chapter 15 of Blaha (2022c) and earlier books by the author. An example of the partitioning appears in Fig. 10.4 below for the Megaverse (Blaha number N = 6).

Note the separation of fermions and internal symmetry groups is independent of coupling constants and particle masses. See Figs. 10.5 and 10.6.

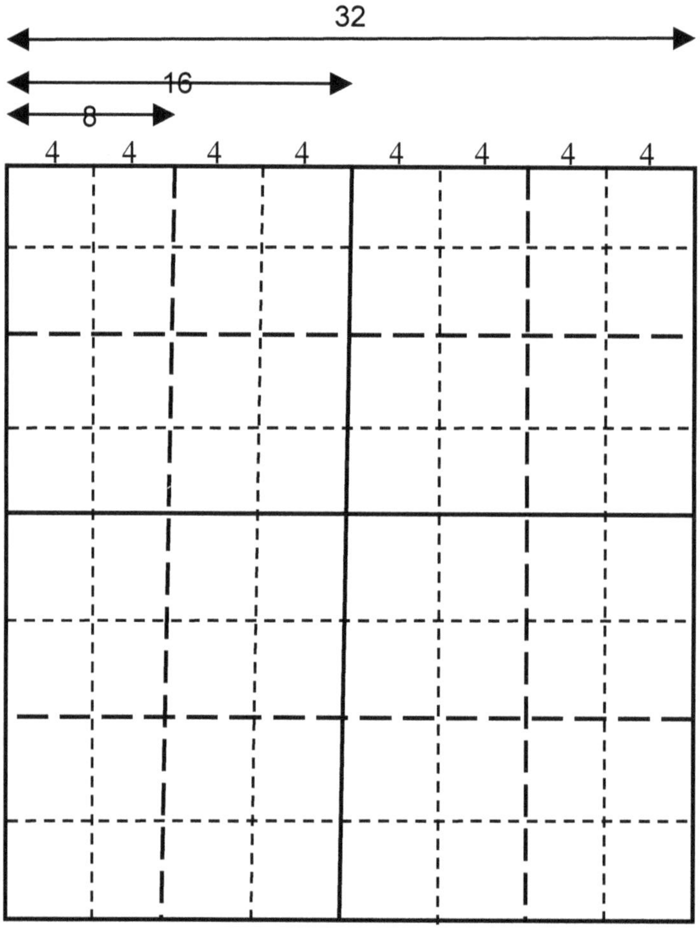

Figure 10.4. The Blaha number N = 6 dimension array quadrisected three times to N = 9 dimension arrays. Quadrisection continues down to N = 9. There are 64 N = 9 space dimension array parts visible above. The horizontal and vertical patterning are the same. This partitioning applies to both fermions and symmetries.

The construction of the set of HyperCosmos spaces leads to a complete specification of internal symmetries, gauge fields, fundamental fermions and Higgs bosons. See Figs. 10.4, 10.5 and 10.6 for the fundamental fermions and symmetries in our universe.[84] The construction is based on the incremental nature of the dimension arrays of the spaces. Each dimension array is effectively composed of four duplicates of the dimension array in the space below it.

The HyperCosmos theory for each space is a Grand Unified Theory (GUT) *ab initio* with a precise specification[85] of particles and symmetries based on mapping dimension array components simply to particles and symmetries.

Each HyperCosmos space has a combined dimension array of space-time dimensions and internal symmetry dimensions. In contrast, the F-Theory approach separates ten coordinates (after compactification) into four space-time coordinates and 6 internal symmetry coordinates, which it then manipulates into particle internal symmetry groups.

	NORMAL				DARK			
	e	q-up	v	q-down	e	q-up	v	q-down
	e	q-up	v	q-down	e	q-up	v	q-down
Layer 1	e	q-up	v	q-down	e	q-up	v	q-down
	e	q-up	v	q-down	e	q-up	v	q-down
	e	q-up	v	q-down	e	q-up	v	q-down
	e	q-up	v	q-down	e	q-up	v	q-down
Layer 2	e	q-up	v	q-down	e	q-up	v	q-down
	e	q-up	v	q-down	e	q-up	v	q-down
	e	q-up	v	q-down	e	q-up	v	q-down
	e	q-up	v	q-down	e	q-up	v	q-down
Layer 3	e	q-up	v	q-down	e	q-up	v	q-down
	e	q-up	v	q-down	e	q-up	v	q-down
	e	q-up	v	q-down	e	q-up	v	q-down
	e	q-up	v	q-down	e	q-up	v	q-down
Layer 4	e	q-up	v	q-down	e	q-up	v	q-down
	e	q-up	v	q-down	e	q-up	v	q-down

Figure 10.5. In QUeST/UST (N = 7) there are four layers of fundamental fermions with each layer containing four generations. The fermions are

[84] These tables may be found in many of our books on UST including Blaha (2018e).

[85] The identification of the group corresponding to each set of dimensions associates each set of dimensions with the dimensions of fundamental group representation dimensions. For example SU(4) has eight real dimensions in its fundamental representation.

arranged in SU(4)-like (possibly broken to SU(3)⊗U(1)) sets of four fermions based on the structures specified in chapter 10 of Blaha (2022c).

NORMAL		DARK	
SU(3)⊗U(1) Generation U(4)	SU(2)⊗U(1)⊗SL(2, C) Layer U(4)	SU(3)⊗U(1) Generation U(4)	SU(2)⊗U(1)⊗SL(2, C) Layer U(4)
SU(3)⊗U(1) Generation U(4)	SU(2)⊗U(1)⊗SL(2, C) Layer U(4)	SU(3)⊗U(1) Generation U(4)	SU(2)⊗U(1)⊗SL(2, C) Layer U(4)
SU(3)⊗U(1) Generation U(4)	SU(2)⊗U(1)⊗SL(2, C) Layer U(4)	SU(3)⊗U(1) Generation U(4)	SU(2)⊗U(1)⊗SL(2, C) Layer U(4)
SU(3)⊗U(1) Generation U(4)	SU(2)⊗U(1)⊗SL(2, C) Layer U(4)	SU(3)⊗U(1) Generation U(4)	SU(2)⊗U(1)⊗SL(2, C) Layer U(4)

Figure 10.6. The sets of symmetries in QUeST/UST and in its N = 7 HyperCosmos space. Note each boxed element has a 32 real dimension representation. This depiction is also evident in QUeST and the UST. The SL(2, **C**) representation has four coordinates.[86]

10.3 Internal Symmetry GUT in F-Theory

F-Theory begins by mapping 12 dimensions to a 10 dimension type IIB Superstring theory by compactifying it on a two-torus. The 10 dimensions are then partitioned into 4 space-time dimensions and 6 internal symmetry dimensions. Then through a series of *ad hoc* assumptions, apparently solely motivated by the goal of having a realistic GUT emerge, a set of 10^{15} possible solutions is found that are consistent with the Standard Model of particles. [87]

It is quite clear that HyperCosmos theory provides a more direct GUT then F-Theory. It has fewer assumptions and selects one theory of elementary particles based on a simple general map from particles and symmetries to dimension arrays. The HyperCosmos approach holds for higher dimension spaces and provides a unified theory in each space.

At present, F-Theory only exists for our universe. The extension to higher dimension universes is problematic since F-Theory doesn't have a dimension independent method to generate GUT equivalents in those spaces/universes. The development of a universal F-type Theory would be interesting.

---

[86] The Lorentz Group $SO^{+}(1, 3)$ is often specified with an SL(2, **C**) representation.

[87] J. J. Heckman, arXiv 1001.0577 (hep-th) (2010), and references therein,.explore the extraction of elementary particle Physics from F-Theory.

11. HyperCosmos and UST Symmetry Splittings

This chapter shows the origin of symmetry splitting is due to the origin of particles and symmetries in Fundamental Reference Frames (chapters 8, 9 and 10), and, particularly, is due to the structure of the dimension arrays derived from the structure of creation/annihilation operators found in chapter 6. We developed this viewpoint originally in Blaha (2021b).

In section 4.3 we developed an interaction for the decay of a universe of space type N to four universes of space type N + 1. This interaction interrelates the ten spaces and their universes. As a consequence the structures within all 10 spaces are interrelated—thus supporting the concept of symmetry splittings presented here for all spaces.

Symmetry splittings implies coupling constant values and particle masses are not the cause the splitting/separation of most symmetries. Exception: The splitting of ElectroWeak-like symmetries is due to the Higgs Mechanism as shown later.

The chapter begins by considering symmetry splitting for our N = 7 universe. It then describes symmetry splitting in general for all 10 spaces and their universes. Then symmetry breakdown due to the Higgs Mechanism is explored. It is also explored in chapter 13 in relation to quark confinement: the Higgs-Confinement Dichotomy.

11.1 Symmetry Splitting in our N = 7 Universe

Symmetry splitting appears to be a result of the structure of the dimension array of $16 \times 16 = 256$ components.

11.1.1 Cayley Number View

If one considers the structure of the N = 7 dimension array, from the view of Cayley numbers, where it is composed of a quadrupling of the N = 8 dimension array of 64 components, which in turn is a quadrupling of the N = 9 dimension array of 16 components, which in turn is a quadrupling of the N = 10 dimension array of 4 components, then one sees the 256 dimension array of our universe is ultimately composed of 4 component parts, replicated into 16 component parts and thence to 64 component parts. This results in the form of the fundamental fermions seen in Fig. 7.2 where the separation into parts is very evident. It also yields the symmetry group separations by fundamental representation dimensions visible in Figs. 7.4 and 7.5.

11.1.2 Fundamental Representation Derivation (FRF)

At a deeper level we can base the symmetry splittings on the generation of symmetries from a Fundamental Reference Frame (FRF) described in chapters 8, 9 and 10.

The 256 component fundamental representation dimension array are generated as replicates from a 16 component array in a FRF by a General Relativistic – Internal

Symmetry transformation[88] as shown in chapters 8 and 9. The 16 component FRF array is structured using a two time dimension separation into subluminal and superluminal parts as shown in chapter 10. Consequently the separation of the 256 component dimension array into symmetry groups as shown earlier gives the QUeST/UST group structure without consideration of symmetry breaking or coupling constants.

We conclude that the splitting of the symmetry groups of our universe into 4 and 8 space-time dimension parts are implicit in the form of the N = 7 HyperCosmos space.

11.2 Symmetry Splitting in a Space N Universe

Symmetry splitting in a universe of the type of HyperCosmos space N is the result of the structure of its dimension array with $d_{dN} = 2^{22-2N}$ elements.

11.2.1 Cayley Number View of Space N

If one considers the structure of the array, from the view of Cayley numbers, we note it is composed of a consecutive quadrupling of spaces for values N' > N in a manner similar to that of section 11.1.1 above.[89]

Thus ultimately the N dimension array is composed of units of 4 and 8 components of space of the N = 10 dimension array of 4 components.

These units assemble into units of 16, then units of 64, and so on to generate a split of symmetries analogous to the N = 7 symmetry group separations by fundamental representation dimensions visible in Figs. 7.4 and 7.5.

11.2.2 Fundamental Representation Derivation (FRF)

Again at a deeper level we can base the symmetry splittings on the generation of symmetries from a Fundamental Reference Frame (FRF) described in chapters 8 and 9.

The $d_{dN} = 2^{22-2N}$ component fundamental representation dimension array can be generated from a 2^{11-N} component vector in a FRF by a General Relativistic – Internal Symmetry transformation[90] as shown in section 8.5 in chapter 8. The generation transformation creates 2^{11-N} distinct replicas of the 2^{11-N} vector's components.

The 2^{11-N} FRF component vector can be factored into 2^{7-N} units of 16 components. Each of these units can be resolved according to superluminal/subliminal separation shown in chapter 10.

Consequently the separation of space N's 2^{22-2N} component dimension array leads to a unique separation of symmetry groups as shown in earlier books without consideration of symmetry breaking or coupling constant values.

We conclude that the splitting of the symmetry groups of a space N universe is implicit in the form of the HyperCosmos space and not due to symmetry breaking or coupling constant values except for ElectroWeak-like symmetry breaking of the SU(2)$\otimes$U(1) .

---

[88] Actually the transformation is one of a set of transformations that transform from one of a set of FRFs.

[89] Fig. 10.1 shows the splitting of an N = 6 universe.

[90] Again the transformation is one of a set of transformations that transform from one of a set of FRFs.

11.3 Scalar Bosons ElectroWeak-like Symmetry Breaking

The ElectroWeak-like *SU(2)*⊗U(1) groups in Fig. 9.4 have symmetry breakdown using the Higgs Mechanism scalar bosons in Fig. 9.2b. We indicate the broken symmetries in Fig. 11.2. Higgs Mechanism breakdowns also appear in the higher dimension SU(2)⊗U(1). Since the higher dimension HyperSpaces are generated effectively by sequences of quadruplings discussed earlier, they also have Higgs Mechanism breakdowns for ElectroWeak-like SU(2)⊗U(1) groups.

11.4 The Standard Model

The preceding developments (plus chapter 12 on interactions) combine to support a conventional Standard Model subsector:

1. Fermion spectrum – The unshaded parts of Fig. 7.3.
2. Symmetries and Gauge vector bosons– Top row of layer 1 of Fig. 7.4.
3. Scalar Bosons – Layer 1 top row of Fig. 9.2b excluding the Dark sector.
4. Symmetry breaking and Symmetry Group Separation – Chapter 9 and this chapter.
5. Interactions – Chapter 12.
6. Gravity and Quantum Gravity – Blaha (2018e). Chapter 14.

Thus the known Standard Model is embodied in the UST and the HyperCosmos for an N = 7 universe.

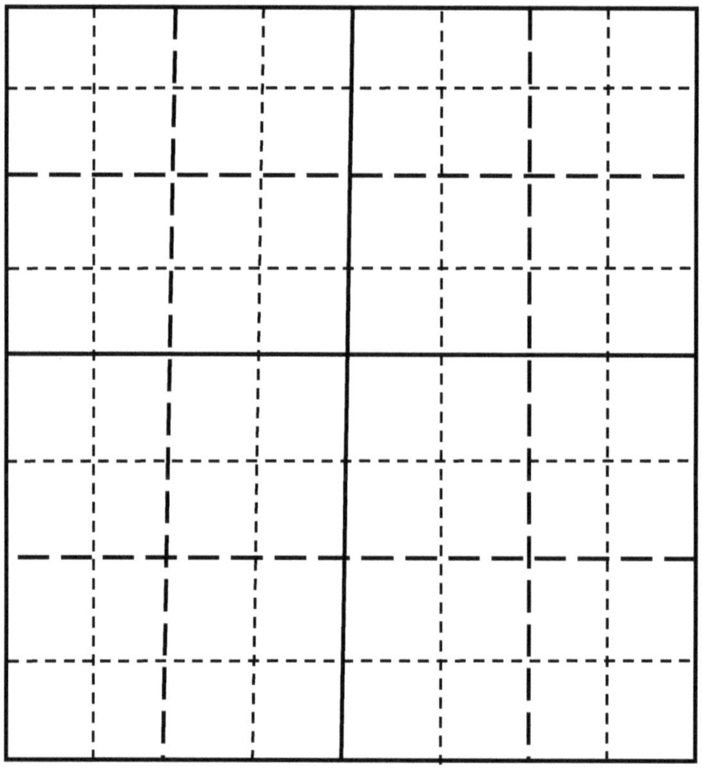

Figure 11.1. The Blaha number N = 6 dimension array quadrisected three times to N = 9 dimension arrays. There are 64 N = 9 space parts visible.

Level	NORMAL		DARK		
1					
	SU(4)	SU(2)⊗U(1)	SU(4)	SU(2)⊗U(1)	(I)
	U(4)	SL(2,**C**)[91]	U(4)	U(2)	(II)
	U(4)		U(4)		(III)
2					
	SU(4)	SU(2)⊗U(1)	SU(4)	SU(2)⊗U(1)	(I)
	U(4)	U(2)	U(4)	U(2)	(II)
	U(4)		U(4)		(III)
3					
	SU(4)	SU(2)⊗U(1)	SU(4)	SU(2)⊗U(1)	(I)
	U(4)	U(2)	U(4)	U(2)	(II)
	U(4)		U(4)		(III)
4					
	SU(4)	SU(2)⊗U(1)	SU(4)	SU(2)⊗U(1)	(I)
	U(4)	U(2)	U(4)	U(2)	(II)
	U(4)		U(4)		(III)

Figure 11.2. The contents of the N = 7 symmetries with ElectroWeak-like SU(2)⊗U(1) broken groups shaded. The separation of symmetries due to su(1, 1) with two times. The separation by two light speeds accounts for the QUeST/UST (and the Standard Model) form of the interaction groups. Line (I) is for the Standard Model sector enhanced to four generations and to a Dark sector. Line. (II) corresponds to the Generation group, and an U(2) Connection Group[92] described earlier except line (II) of layer 1 which has a space-time group SL(2, **C**). Line (III) corresponds to the Layer Group also described earlier.

11.5 Evidence for HyperCosmos Structure and Two Time Coordinate Splitting

The form of HyperCosmos space structure which bases dimension array structure on the quadrupling process of the dimension arrays from space to space gives the structuring of groups seen in the Unified SuperStandard Theory (UST) based on the "magic" numbers 4, 8, 16, 32, and 64. The form of the two times separation of groups into subluminal and superluminal sectors gives the SU(2)⊗U(1)⊗SU(3) structuring of the Standard Model symmetry groups and the UST. It also gives the differentiation between the Higgs Mechanism, which is viewed as superluminal, and Strong Interaction Confinement, which is viewed as subluminal and described in chapter 13.

Together these experimentally known points are explained, and provide strong support, for the HyperCosmos and Fundamental Reference Frame based unification.

---

[91] $SO^+(1, 3)$.

[92] See Appendix 7-B.

12. N = 7 Universe Interactions

12.1 Dynamics and Unification

The unification of General Relativity and Quantum Theory is evident in the HyperCosmos at several levels. The complete set of symmetry groups, including the space-time group, is contained in each space. The dimension array with $2^{r/2+2} \times 2^{r/2+2} = 2^{r+4}$ real dimensions is the irreducible representation of a SU(2^{r+3}) symmetry group that is broken into the set of subgroups for internal symmetries and space-time for each HyperCosmos space. We view the SU(2^{r+3}) group as the initial group at the point of creation – the Big Bang, which has extremely large energy concentrated at one point.[93]

The explicit unity of the symmetry groups before separation into subgroups shows unification of the symmetry groups as one expects in Grand Unified Theories (GUTS) thus fulfilling one type of unification.

Another, perhaps deeper, form of unification stems from the CASe groups of GiFT. We remember the CASe groups were the source of the fundamental symmetry group dimension array. The CASe groups were defined based on the need for General Relativistic transformations of sets of creation/annihilation operators between static and non-static coordinate systems. General Relativity is thus at the heart of Quantum Theory. The requirement of covariance in both dynamic equations and in creation/annihilation operator expressions completes the unification program for Quantum Theory and General Relativity.

CASe is required by General Relativistic – Internal Symmetry transformations of Quantum creation/annihilation operators to non-static coordinate systems. The dimension arrays of the symmetry groups follow from CASe and give an initial unified symmetry group for each HyperCosmos space. The Riemann-Christoffel curvature tensor embodies all the interactions including Gravitation and leads to a complete unified dynamics.

The unification of Quantum Theory and General Relativity is completed by the combination of space-time dimensions and internal symmetry representation dimensions in a dimension array in each HyperCosmos space.

An N = 7 universe has the internal symmetry groups in layer 1

$$[SU(4)\otimes SU(2)\otimes U(1)\otimes U(4)\otimes U(4)]^2\otimes U(2) \qquad (12.1)$$

and in layers 2 , 3, and 4:

$$[SU(4)\otimes SU(2)\otimes U(1)\otimes U(4)\otimes U(4)\otimes U(2)]^2 \qquad (12.2)$$

according to Fig. 12.1. The total internal symmetry is

---

[93] Blaha (2021d) shows the Big Bang and subsequent expansion can be viewed as analogous to vacuum polarization. It shows free field behavior at the Big Bang.

$$[SU(4)\otimes SU(2)\otimes U(1)\otimes U(4)\otimes U(4)]^8\otimes U(2)^7 \qquad (12.3)$$

The group parts are

NORMAL	DARK	CONNECTION
$[SU(4)\otimes SU(2)\otimes U(1)\otimes U(4)\otimes U(4)]^4$	$\otimes[SU(4)\otimes SU(2)\otimes U(1)\otimes U(4)\otimes U(4)]^4$	$\otimes U(2)^7$

(12.4)

Level	NORMAL		DARK		
1					
	SU(4)	SU(2)⊗U(1)	SU(4)	SU(2)⊗U(1)	(I)
	U(4) Generation	SL(2,**C**)	U(4) Generation	U(2)	(II)
	U(4) Layer		U(4) Layer		(III)
2					
	SU(4)	SU(2)⊗U(1)	SU(4)	SU(2)⊗U(1)	(I)
	U(4) Generation	U(2)	U(4) Generation	U(2)	(II)
	U(4) Layer		U(4) Layer		(III)
3					
	SU(4)	SU(2)⊗U(1)	SU(4)	SU(2)⊗U(1)	(I)
	U(4) Generation	U(2)	U(4) Generation	U(2)	(II)
	U(4) Layer		U(4) Layer		(III)
4					
	SU(4)	SU(2)⊗U(1)	SU(4)	SU(2)⊗U(1)	(I)
	U(4) Generation	U(2)	U(4) Generation	U(2)	(II)
	U(4) Layer		U(4) Layer		(III)

Figure 12.1. The contents of the N = 7 symmetries with ElectroWeak-like SU(2)⊗U(1) broken groups shaded from Fig. 11.2. Line (I) is for the Standard Model sector enhanced to four generations and to a Dark sector. Line. (II) corresponds to the Generation group, and an U(2) Connection Group[94] described earlier except line (II) of layer 1 which has a space-time group SL(2, **C**). Line (III) corresponds to the Layer Group also described earlier.

The PseudoQuantum formulation vector gauge fields are[95]

NORMAL TYPE 1 PSEUDOQUANTUM

$$F_{SU(4)}{}^{1}{}_{\varkappa\mu} = \Sigma\ \{\partial A_{SU(4)}{}^{1}{}_{\mu}/\partial x^{\varkappa} - \partial A_{SU(4)}{}^{1}{}_{\varkappa}/\partial x^{\mu} + ig_1[A_{SU(4)}{}^{1}{}_{\varkappa}, A_{SU(4)}{}^{1}{}_{\mu}]\ \} \qquad (12.5)$$

$$F_{W}{}^{1}{}_{\varkappa\mu} = \Sigma\ \{\partial W^{1}{}_{\mu}/\partial x^{\varkappa} - \partial W^{1}{}_{\varkappa}/\partial x^{\mu} + ig_2[W^{1}{}_{\varkappa}, W^{1}{}_{\mu}]\ \}$$

---

[94] See Appendix 7-B.

[95] Stephen Blaha, "New Framework for Gauge Field Theories", IL Nuovo Cimento **49A**, 113 (1979). It describes the PseudoQuantization of gauge field theories for the purposes of defining higher derivative field theories and for use in the quantization of fields in universes and the Megaverse. Also see Blaha (2017d).

$$F_{E}{}^{1}{}_{\varkappa\mu} = \Sigma\ \{\partial A_{E}{}^{1}{}_{\mu}/\partial x^{\varkappa} - \partial A_{E}{}^{1}{}_{\varkappa}/\partial x^{\mu}\ \}$$
$$F_{U}{}^{1}{}_{\varkappa\mu} = \Sigma\ \{\partial U^{1}{}_{\mu}/\partial x^{\varkappa} - \partial U^{1}{}_{\varkappa}/\partial x^{\mu} + ig_{6}[U^{1}{}_{\varkappa}, U^{1}{}_{\mu}]\ \}$$
$$F_{V}{}^{1}{}_{\varkappa\mu} = \Sigma\ \{\partial V^{1}{}_{\mu}/\partial x^{\varkappa} - \partial V^{1}{}_{\varkappa}/\partial x^{\mu} + ig_{7}[V^{1}{}_{\varkappa}, V^{1}{}_{\mu}]\ \}$$

DARK TYPE 1 PSEUDOQUANTUM

$$F_{DSU(4)}{}^{1}{}_{\varkappa\mu} = \Sigma\ \{\partial A_{DSU(4)}{}^{1}{}_{\mu}/\partial x^{\varkappa} - \partial A_{DSU(4)}{}^{1}{}_{\varkappa}/\partial x^{\mu} + ig_{1}[A_{DSU(4)}{}^{1}{}_{\varkappa}, A_{DSU(4)}{}^{1}{}_{\mu}]\ \}$$
$$F_{DW}{}^{1}{}_{\varkappa\mu} = \Sigma\ \{\partial W_{D}{}^{1}{}_{\mu}/\partial x^{\varkappa} - \partial W_{D}{}^{1}{}_{\varkappa}/\partial x^{\mu} + ig_{4}[W_{D}{}^{1}{}_{\varkappa}, W_{D}{}^{1}{}_{\mu}]\ \}$$
$$F_{DE}{}^{1}{}_{\varkappa\mu} = \Sigma\ \{\partial A_{DE}{}^{1}{}_{\mu}/\partial x^{\varkappa} - \partial A_{DE}{}^{1}{}_{\varkappa}/\partial x^{\mu}\}$$
$$F_{DU}{}^{1}{}_{\varkappa\mu} = \Sigma\ \{\partial U_{D}{}^{1}{}_{\mu}/\partial x^{\varkappa} - \partial U_{D}{}^{1}{}_{\varkappa}/\partial x^{\mu} + ig_{6}[U_{D}{}^{1}{}_{\varkappa}, U_{D}{}^{1}{}_{\mu}]\ \}$$
$$F_{DV}{}^{1}{}_{\varkappa\mu} = \Sigma\ \{\partial V_{D}{}^{1}{}_{\mu}/\partial x^{\varkappa} - \partial V_{D}{}^{1}{}_{\varkappa}/\partial x^{\mu} + ig_{7}[V_{D}{}^{1}{}_{\varkappa}, V_{D}{}^{1}{}_{\mu}]\ \}$$

CONNECTION GROUPS TYPE 1 PSEUDOQUANTUM

$$F_{Ca}{}^{1}{}_{\varkappa\mu} = \Sigma\ \{\partial C_{a}{}^{1}{}_{\mu}/\partial x^{\varkappa} - \partial C_{a}{}^{1}{}_{\varkappa}/\partial x^{\mu} + ig_{c}[C_{a}{}^{1}{}_{\varkappa}, C_{a}{}^{1}{}_{\mu}]\ \} \quad \text{for } a = 1,2, \ldots, 7$$

with TYPE 1 spinor potential

$$F_{B}{}^{1}{}_{\varkappa\mu} = \partial B^{1}{}_{\mu}/\partial x^{\varkappa} - \partial B^{1}{}_{\varkappa}/\partial x^{\mu} + i[B^{1}{}_{\varkappa}, B^{1}{}_{\mu}]$$

NORMAL TYPE 2 PSEUDOQUANTUM

$$F_{SU(4)}{}^{2}{}_{\varkappa\mu} = \Sigma\ \{\partial A_{SU(4)}{}^{2}{}_{\mu}/\partial x^{\varkappa} - \partial A_{SU(4)}{}^{2}{}_{\varkappa}/\partial x^{\mu} + ig_{1}[A_{SU(4)}{}^{2}{}_{\varkappa}, A_{SU(4)}{}^{2}{}_{\mu}] + $$
$$+ ig_{1}[A_{SU(4)}{}^{1}{}_{\varkappa}, A_{SU(4)}{}^{2}{}_{\mu}] + ig_{1}[A_{SU(4)}{}^{2}{}_{\varkappa}, A_{SU(4)}{}^{1}{}_{\mu}]\}$$
$$F_{W}{}^{2}{}_{\varkappa\mu} = \Sigma\ \{\partial W^{2}{}_{\mu}/\partial x^{\varkappa} - \partial W^{2}{}_{\varkappa}/\partial x^{\mu} + ig_{2}[W^{2}{}_{\varkappa}, W^{2}{}_{\mu}] + ig_{2}[W^{1}{}_{\varkappa}, W^{2}{}_{\mu}] + ig_{2}[W^{2}{}_{\varkappa}, W^{1}{}_{\mu}]\ \}$$
$$F_{E}{}^{2}{}_{\varkappa\mu} = \Sigma\ \{\partial A_{E}{}^{2}{}_{\mu}/\partial x^{\varkappa} - \partial A_{E}{}^{2}{}_{\varkappa}/\partial x^{\mu}\}$$
$$F_{U}{}^{2}{}_{\varkappa\mu} = \Sigma\ \{\partial U^{2}{}_{\mu}/\partial x^{\varkappa} - \partial U^{2}{}_{\varkappa}/\partial x^{\mu} + ig_{2}[U^{2}{}_{\varkappa}, U^{2}{}_{\mu}] + ig_{6}[U^{1}{}_{\varkappa}, U^{2}{}_{\mu}] + ig_{6}[U^{2}{}_{\varkappa}, U^{1}{}_{\mu}]\ \}$$
$$F_{V}{}^{2}{}_{\varkappa\mu} = \Sigma\ \{\partial V^{2}{}_{\mu}/\partial x^{\varkappa} - \partial V^{2}{}_{\varkappa}/\partial x^{\mu} + ig_{2}[V^{2}{}_{\varkappa}, V^{2}{}_{\mu}] + ig_{7}[V^{1}{}_{\varkappa}, V^{2}{}_{\mu}] + ig_{7}[V^{2}{}_{\varkappa}, V^{1}{}_{\mu}]\ \}$$
$$F_{W}{}^{2}{}_{\varkappa\mu} = \Sigma\ \{\partial W^{2}{}_{\mu}/\partial x^{\varkappa} - \partial W^{2}{}_{\varkappa}/\partial x^{\mu} + ig_{2}[W^{2}{}_{\varkappa}, W^{2}{}_{\mu}] + ig_{2}[W^{1}{}_{\varkappa}, W^{2}{}_{\mu}] + ig_{2}[W^{2}{}_{\varkappa}, W^{1}{}_{\mu}]\ \}$$

CONNECTION GROUPS TYPE 2 PSEUDOQUANTUM

$$F_{Ca}{}^{2}{}_{\varkappa\mu} = \Sigma\{\partial C_{a}{}^{2}{}_{\mu}/\partial x^{\varkappa} - \partial\ C_{a}{}^{2}{}_{\varkappa}/\partial x^{\mu} + ig_{2}[C_{a}{}^{2}{}_{\varkappa}, C_{a}{}^{2}{}_{\mu}] + ig_{2}[C_{a}{}^{1}{}_{\varkappa}, C_{a}{}^{2}{}_{\mu}] + ig_{2}[C_{a}{}^{2}{}_{\varkappa}, C_{a}{}^{1}{}_{\mu}]\ \}$$
$$\text{for } a = 1,2, \ldots, 7.$$

with TYPE 2 spinor field

$$F_{B}{}^{2}{}_{\varkappa\mu} = \partial B^{2}{}_{\mu}/\partial x^{\varkappa} - \partial B^{2}{}_{\varkappa}/\partial x^{\mu} + i[B^{2}{}_{\mu}, B^{2}{}_{\varkappa}] + i[B^{1}{}_{\mu}, B^{2}{}_{\varkappa}] + i[B^{2}{}_{\mu}, B^{1}{}_{\varkappa}]$$

DARK TYPE 2 PSEUDOQUANTUM

$$F_{DSU(4)}{}^{2}{}_{\varkappa\mu} = \Sigma\ \{\partial A_{DSU(4)}{}^{2}{}_{\mu}/\partial x^{\varkappa} - \partial A_{DSU(4)}{}^{2}{}_{\varkappa}/\partial x^{\mu} + ig_{1}[A_{DSU(4)}{}^{2}{}_{\varkappa}, A_{DSU(4)}{}^{2}{}_{\mu}] + $$
$$+ ig_{1}[A_{DSU(4)}{}^{1}{}_{\varkappa}, A_{DSU(4)}{}^{2}{}_{\mu}] + ig_{1}[A_{DSU(4)}{}^{2}{}_{\varkappa}, A_{DSU(4)}{}^{1}{}_{\mu}]\}$$

$$F_{DW}{}^{2}{}_{\varkappa\mu} = \Sigma\ \{\partial W_{D}{}^{2}{}_{\mu}/\partial x^{\varkappa} - \partial W_{D}{}^{2}{}_{\varkappa}/\partial x^{\mu} + ig_{4}[W_{D}{}^{2}{}_{\varkappa}, W_{D}{}^{2}{}_{\mu}] + ig_{4}[W_{D}{}^{1}{}_{\varkappa}, W_{D}{}^{2}{}_{\mu}] + $$
$$+ ig_{4}[W_{D}{}^{2}{}_{\varkappa}, W_{D}{}^{1}{}_{\mu}]\}$$

$$F_{DE}{}^{2}{}_{\varkappa\mu} = \Sigma\ \{\partial A_{DE}{}^{2}{}_{\mu}/\partial x^{\varkappa} - \partial A_{DE}{}^{2}{}_{\varkappa}/\partial x^{\mu}\}$$
$$F_{DU}{}^{2}{}_{\varkappa\mu} = \Sigma\ \{\partial U_{D}{}^{2}{}_{\mu}/\partial x^{\varkappa} - \partial U_{D}{}^{2}{}_{\varkappa}/\partial x^{\mu} + ig_6[U_{D}{}^{2}{}_{\varkappa}, U_{D}{}^{2}{}_{\mu}] + ig_6[U_{D}{}^{1}{}_{\varkappa}, U_{D}{}^{2}{}_{\mu}] + ig_6[U_{D}{}^{2}{}_{\varkappa}, U_{D}{}^{1}{}_{\mu}]\}$$
$$F_{DV}{}^{2}{}_{\varkappa\mu} = \Sigma\ \{\partial V_{D}{}^{2}{}_{\mu}/\partial x^{\varkappa} - \partial V_{D}{}^{2}{}_{\varkappa}/\partial x^{\mu} + ig_7[V_{D}{}^{2}{}_{\varkappa}, V_{D}{}^{2}{}_{\mu}] + ig_7[V_{D}{}^{1}{}_{\varkappa}, V_{D}{}^{2}{}_{\mu}] + ig_7[V_{D}{}^{2}{}_{\varkappa}, V_{D}{}^{1}{}_{\mu}]\}$$

Using the above definitions the PseudoQuantum covariant derivative of a 4-vector is

$$\begin{aligned} D_\nu Z_\mu &= (\partial_\nu + iF_\nu)Z_\mu - H^\sigma{}_{\nu\mu}Z_\sigma \\ &= [g^\sigma{}_\mu\partial_\nu + ig^\sigma{}_\mu F_\nu - H^\sigma{}_{\nu\mu}]Z_\sigma \\ &= [g^\sigma{}_\mu\partial_\nu + iD^\sigma{}_{\mu\nu}]Z_\sigma \end{aligned} \tag{12.6}$$

where[96]

$$F^\mu = \Sigma_a\ [F_a{}^{1\mu}(x) + \mathbf{F}_a{}^{2\mu}(x) + F_{Da}{}^{1\mu}(x) + \mathbf{F}_{Da}{}^{2\mu}(x)\ ] \tag{12.7}$$

with a summation over all entries in the parts of eq. 12.5 and

$$\begin{aligned} H^\sigma{}_{\nu\mu} &= \Gamma_{GR}{}^{1\sigma}{}_{\nu\mu} + \Gamma_{GR}{}^{2\sigma}{}_{\nu\mu} \\ D^\sigma{}_{\mu\nu} &= g^\sigma{}_\mu F_\nu + iH^\sigma{}_{\nu\mu} \end{aligned} \tag{12.8}$$

for PseudoQuantum type 1 and 2 $\Gamma_{GR}{}^{1\sigma}{}_{\nu\mu}$ and $\Gamma_{GR}{}^{2\sigma}{}_{\nu\mu}$ respectively.

The curvature tensor applied to a 4-vector is

$$\begin{aligned} R'^\beta{}_{\sigma\nu\mu}Z_\beta = {} & g^\alpha{}_\mu(\partial_\nu + iF_\nu)g^\beta{}_\sigma(\partial_\alpha + iF_\alpha)Z_\beta - H^\alpha{}_{\mu\nu}g^\beta{}_\sigma(\partial_\alpha + iF_\alpha)Z_\beta + \\ & + H^\alpha{}_{\mu\nu}H^\beta{}_{\sigma\alpha}Z_\beta - g^\alpha{}_\mu(\partial_\nu + iF_\nu)H^\beta{}_{\sigma\alpha}Z_\beta - H^\gamma{}_{\nu\sigma}\{g^\alpha{}_\gamma(\partial_\mu + iF_\mu)Z_\alpha - H^\alpha{}_{\gamma\mu}Z_\alpha\} - \\ & - \{\mu\leftrightarrow\nu\} \end{aligned} \tag{12.9}$$

$$= ig^\beta{}_\sigma(\partial_\nu F_\mu - \partial_\mu F_\nu\ - i[F_\nu, F_\mu])Z_\beta + (\partial_\mu H^\beta{}_{\sigma\nu} - \partial_\nu H^\beta{}_{\sigma\mu} + H^\gamma{}_{\nu\sigma}H^\beta{}_{\gamma\mu} - H^\gamma{}_{\mu\sigma}H^\beta{}_{\gamma\nu})Z_\beta$$

$$\begin{aligned} = {} & ig^\beta{}_\sigma(F_{E}{}^{1}{}_{\nu\mu} + F_{E}{}^{2}{}_{\nu\mu} + F_{W}{}^{1}{}_{\nu\mu} + F_{W}{}^{2}{}_{\nu\mu} + F_{SU(4)}{}^{1}{}_{\nu\mu} + F_{SU(4)}{}^{2}{}_{\nu\mu} + F_{U}{}^{1}{}_{\nu\mu} + F_{U}{}^{2}{}_{\nu\mu} + F_{V}{}^{1}{}_{\nu\mu} + F_{V}{}^{2}{}_{\nu\mu} \\ & + F_{DE}{}^{1}{}_{\nu\mu} + F_{DE}{}^{2}{}_{\nu\mu} + F_{DW}{}^{1}{}_{\nu\mu} + F_{DW}{}^{2}{}_{\nu\mu} + F_{DSU(4)}{}^{1}{}_{\nu\mu} + F_{DSU(4)}{}^{2}{}_{\nu\mu} + F_{DU}{}^{1}{}_{\nu\mu} + F_{DU}{}^{2}{}_{\nu\mu} + F_{DV}{}^{1}{}_{\nu\mu} + \\ & + F_{DV}{}^{2}{}_{\nu\mu} + F_{B}{}^{1}{}_{\nu\mu} + F_{B}{}^{2}{}_{\nu\mu} + \Sigma_a(F_{Ca}{}^{1}{}_{\nu\mu} + F_{Ca}{}^{2}{}_{\nu\mu}))Z_\beta + (\partial_\mu H^\beta{}_{\sigma\nu} - \partial_\nu H^\beta{}_{\sigma\mu} + H^\gamma{}_{\nu\sigma}H^\beta{}_{\gamma\mu} - \\ & - H^\gamma{}_{\mu\sigma}H^\beta{}_{\gamma\nu})Z_\beta \end{aligned}$$

where all $F_{\ldots}{}^{1}{}_{\nu\mu}$ and $F_{\ldots}{}^{2}{}_{\nu\mu}$ terms have summations over the four layers (see below) except the terms $F_{B}{}^{1}{}_{\nu\mu} + F_{B}{}^{2}{}_{\nu\mu}$, and $\Sigma_a(F_{Ca}{}^{1}{}_{\nu\mu} + F_{Ca}{}^{2}{}_{\nu\mu})$, and where

$$\begin{aligned} R'_G{}^\beta{}_{\sigma\nu\mu} &= \partial_\mu H^\beta{}_{\sigma\nu} - \partial_\nu H^\beta{}_{\sigma\mu} + H^\gamma{}_{\nu\sigma}H^\beta{}_{\gamma\mu} - H^\gamma{}_{\mu\sigma}H^\beta{}_{\gamma\nu} \\ &= \partial_\mu H^{1\beta}{}_{\sigma\nu} - \partial_\nu H^{1\beta}{}_{\sigma\mu} + H^{1\gamma}{}_{\nu\sigma}H^{1\beta}{}_{\gamma\mu} - H^{1\gamma}{}_{\mu\sigma}H^{1\beta}{}_{\gamma\nu} + \partial_\mu H^{2\beta}{}_{\sigma\nu} - \partial_\nu H^{2\beta}{}_{\sigma\mu} + \end{aligned} \tag{12.10}$$

---

[96] We will omit the insertion of the spinor coupling constants of $B^{1\mu}$ and $B^{2\mu}$ in eq. 12.7 in the interests of simplifying expressions.

$$+ H^{2\gamma}{}_{\nu\sigma}H^{2\beta}{}_{\gamma\mu} - H^{2\gamma}{}_{\mu\sigma}H^{2\beta}{}_{\gamma\nu} + H^{1\gamma}{}_{\nu\sigma}H^{2\beta}{}_{\gamma\mu} - H^{1\gamma}{}_{\mu\sigma}H^{2\beta}{}_{\gamma\nu} + H^{2\gamma}{}_{\nu\sigma}H^{1\beta}{}_{\gamma\mu} - \Gamma^{2\gamma}{}_{\mu\sigma}\Gamma^{\beta}{}_{\gamma\nu}$$
$$= R^{1\beta}{}_{\sigma\nu\mu} + R^{2\beta}{}_{\sigma\nu\mu}$$

with

$$\begin{aligned}
H^{\beta}{}_{\sigma\nu\mu} &= \partial_\mu H^{\beta}{}_{\sigma\nu} - \partial_\nu H^{\beta}{}_{\sigma\mu} + H^{\gamma}{}_{\nu\sigma}H^{\beta}{}_{\gamma\mu} - H^{\gamma}{}_{\mu\sigma}H^{\beta}{}_{\gamma\nu} \\
R^{1\beta}{}_{\sigma\nu\mu} &= \partial_\mu H^{1\beta}{}_{\sigma\nu} - \partial_\nu H^{1\beta}{}_{\sigma\mu} + H^{1\gamma}{}_{\nu\sigma}H^{1\beta}{}_{\gamma\mu} - H^{1\gamma}{}_{\mu\sigma}H^{1\beta}{}_{\gamma\nu} \\
R^{2\beta}{}_{\sigma\nu\mu p} &= \partial_\mu H^{2\beta}{}_{\sigma\nu} - \partial_\nu H^{2\beta}{}_{\sigma\mu} + H^{2\gamma}{}_{\nu\sigma}H^{2\beta}{}_{\gamma\mu} - H^{2\gamma}{}_{\mu\sigma}H^{2\beta}{}_{\gamma\nu} + \\
&\quad + H^{1\gamma}{}_{\nu\sigma}H^{2\beta}{}_{\gamma\mu} - H^{1\gamma}{}_{\mu\sigma}H^{2\beta}{}_{\gamma\nu} + H^{2\gamma}{}_{\nu\sigma}H^{1\beta}{}_{\gamma\mu} - H^{2\gamma}{}_{\mu\sigma}H^{1\beta}{}_{\gamma\nu}
\end{aligned} \tag{12.11}$$

and

$$\begin{aligned}
H^{1\sigma}{}_{\nu\mu} &= \Gamma_{GR}{}^{1\sigma}{}_{\nu\mu} \\
H^{2\sigma}{}_{\nu\mu} &= \Gamma_{GR}{}^{2\sigma}{}_{\nu\mu}
\end{aligned} \tag{12.12}$$

with summations over the four layers.

For later use in defining a lagrangian we define

$$\begin{aligned}
R'^{\beta}{}_{\sigma\nu\mu} &= R'_E{}^{1\beta}{}_{\sigma\nu\mu} + R'_E{}^{2\beta}{}_{\sigma\nu\mu} + R'_{SU(2)}{}^{1\beta}{}_{\sigma\nu\mu} + R'_{SU(2)}{}^{2\beta}{}_{\sigma\nu\mu} + R'_{SU(4)}{}^{1\beta}{}_{\sigma\nu\mu} + R'_{SU(4)}{}^{2\beta}{}_{\sigma\nu\mu} + \\
&+ R'_U{}^{1\beta}{}_{\sigma\nu\mu} + R'_U{}^{2\beta}{}_{\sigma\nu\mu} + R'_V{}^{1\beta}{}_{\sigma\nu\mu} + R'_V{}^{2\beta}{}_{\sigma\nu\mu} + \\
&+ R'_{DE}{}^{1\beta}{}_{\sigma\nu\mu} + R'_{DE}{}^{2\beta}{}_{\sigma\nu\mu} + R'_{DSU(2)}{}^{1\beta}{}_{\sigma\nu\mu} + R'_{DSU(2)}{}^{2\beta}{}_{\sigma\nu\mu} + R'_{DSU(4)}{}^{1\beta}{}_{\sigma\nu\mu} + R'_{DSU(4)}{}^{2\beta}{}_{\sigma\nu\mu} + \\
&+ R'_{DU}{}^{1\beta}{}_{\sigma\nu\mu} + R'_{DU}{}^{2\beta}{}_{\sigma\nu\mu} + R'_{DV}{}^{1\beta}{}_{\sigma\nu\mu} + R'_{DV}{}^{2\beta}{}_{\sigma\nu\mu} + \Sigma_a(R'_{Ca^1}{}^{\beta}{}_{\sigma\nu\mu} + R'_{Ca^2}{}^{\beta}{}_{\sigma\nu\mu}) + R'_B{}^{1\beta}{}_{\sigma\nu\mu} + \\
&+ R'_B{}^{2\beta}{}_{\sigma\nu\mu} + R^{1\beta}{}_{\sigma\nu\mu} + R^{2\beta}{}_{\sigma\nu\mu}
\end{aligned} \tag{12.13}$$

where

$$\begin{aligned}
R'_E{}^{1\beta}{}_{\sigma\nu\mu} &= ig^{\beta}{}_{\sigma}F_E{}^{1}{}_{\nu\mu} \\
R'_E{}^{2\beta}{}_{\sigma\nu\mu} &= ig^{\beta}{}_{\sigma}F_E{}^{2}{}_{\nu\mu}
\end{aligned} \tag{12.14}$$

$$R'_{DE}{}^{1\beta}{}_{\sigma\nu\mu} = ig^{\beta}{}_{\sigma}F_{DE}{}^{1}{}_{\nu\mu}$$
$$R'_{DE}{}^{2\beta}{}_{\sigma\nu\mu} = ig^{\beta}{}_{\sigma}F_{DE}{}^{2}{}_{\nu\mu}$$

$$R'_{SU(2)}{}^{1\beta}{}_{\sigma\nu\mu} = ig^{\beta}{}_{\sigma}F_W{}^{1}{}_{\nu\mu}$$
$$R'_{SU(2)}{}^{2\beta}{}_{\sigma\nu\mu} = ig^{\beta}{}_{\sigma}F_W{}^{2}{}_{\nu\mu}$$

$$R'_{DSU(2)}{}^{1\beta}{}_{\sigma\nu\mu} = ig^{\beta}{}_{\sigma}F_{DW}{}^{1}{}_{\nu\mu}$$
$$R'_{DSU(2)}{}^{2\beta}{}_{\sigma\nu\mu} = ig^{\beta}{}_{\sigma}F_{DW}{}^{2}{}_{\nu\mu}$$

$$R'_{SU(4)}{}^{1\beta}{}_{\sigma\nu\mu} = ig^{\beta}{}_{\sigma}F_{SU(4)}{}^{1}{}_{\nu\mu}$$
$$R'_{SU(4)}{}^{2\beta}{}_{\sigma\nu\mu} = ig^{\beta}{}_{\sigma}F_{SU(4)}{}^{2}{}_{\nu\mu}$$

$$R'_{DSU(4)}{}^{1\beta}{}_{\sigma\nu\mu} = ig^{\beta}{}_{\sigma}F_{DSU(4)}{}^{1}{}_{\nu\mu}$$
$$R'_{DSU(4)}{}^{2\beta}{}_{\sigma\nu\mu} = ig^{\beta}{}_{\sigma}F_{DSU(4)}{}^{2}{}_{\nu\mu}$$

$$R'_U{}^{1\beta}{}_{\sigma\nu\mu} = ig^{\beta}{}_{\sigma}F_U{}^{1}{}_{\nu\mu}$$
$$R'_U{}^{2\beta}{}_{\sigma\nu\mu} = ig^{\beta}{}_{\sigma}F_U{}^{2}{}_{\nu\mu}$$

$$R'_{DU}{}^{1\beta}{}_{\sigma\nu\mu} = ig^{\beta}{}_{\sigma}F_{DU}{}^{1}{}_{\nu\mu}$$
$$R'_{DU}{}^{2\beta}{}_{\sigma\nu\mu} = ig^{\beta}{}_{\sigma}F_{DU}{}^{2}{}_{\nu\mu}$$

$$R'_{V}{}^{1\beta}{}_{\sigma\nu\mu} = ig^{\beta}{}_{\sigma}F_{V}{}^{1}{}_{\nu\mu}$$
$$R'_{V}{}^{2\beta}{}_{\sigma\nu\mu} = ig^{\beta}{}_{\sigma}F_{V}{}^{2}{}_{\nu\mu}$$

$$R'_{DV}{}^{1\beta}{}_{\sigma\nu\mu} = ig^{\beta}{}_{\sigma}F_{DV}{}^{1}{}_{\nu\mu}$$
$$R'_{DV}{}^{2\beta}{}_{\sigma\nu\mu} = ig^{\beta}{}_{\sigma}F_{DV}{}^{2}{}_{\nu\mu}$$

$$R'_{Ca}{}^{1\beta}{}_{\sigma\nu\mu} = ig^{\beta}{}_{\sigma}F_{Ca}{}^{1}{}_{\nu\mu}$$
$$R'_{Ca}{}^{2\beta}{}_{\sigma\nu\mu} = ig^{\beta}{}_{\sigma}F_{Ca}{}^{2}{}_{\nu\mu}$$

$$R'_{B}{}^{1\beta}{}_{\sigma\nu\mu} = ig^{\beta}{}_{\sigma}B^{1}{}_{\nu\mu}$$
$$R'_{B}{}^{2\beta}{}_{\sigma\nu\mu} = ig^{\beta}{}_{\sigma}B^{2}{}_{\nu\mu}$$

with the further definition of $R''^{1}{}_{\sigma\mu}$ and $R''^{2}{}_{\sigma\mu}$:

$$R''^{1}{}_{\sigma\mu} = R'_{SU(3)}{}^{1}{}_{\sigma\mu} + R^{1}{}_{\sigma\mu}$$
$$R''^{2}{}_{\sigma\mu} = R'_{SU(3)}{}^{2}{}_{\sigma\mu} + R^{2}{}_{\sigma\mu} \tag{22.15}$$

The total Ricci tensor is

$$\begin{aligned} R'_{\sigma_\mu} &= R'^{\beta}{}_{\sigma\beta\mu} \qquad (12.16)\\ &= iF_{E}{}^{1}{}_{\sigma\mu} + iF_{E}{}^{2}{}_{\sigma\mu} + iF_{W}{}^{1}{}_{\sigma\mu} + iF_{W}{}^{2}{}_{\sigma\mu} + iF_{DE}{}^{1}{}_{\sigma\mu} + iF_{DE}{}^{2}{}_{\sigma\mu} + iF_{DW}{}^{1}{}_{\sigma\mu} + iF_{DW}{}^{2}{}_{\sigma\mu} + iF_{SU(4)}{}^{1}{}_{\sigma\mu} + iF_{SU(4)}{}^{2}{}_{\sigma\mu} +\\ &\quad + iF_{DSU(4)}{}^{1}{}_{\sigma\mu} + iF_{DSU(4)}{}^{2}{}_{\sigma\mu} + iF_{U}{}^{1}{}_{\sigma\mu} + iF_{U}{}^{2}{}_{\sigma\mu} + iF_{V}{}^{1}{}_{\sigma\mu} + iF_{V}{}^{2}{}_{\sigma\mu} + iF_{DU}{}^{1}{}_{\sigma\mu} + iF_{DU}{}^{2}{}_{\sigma\mu} + iF_{DV}{}^{1}{}_{\sigma\mu} +\\ &\quad + iF_{DV}{}^{2}{}_{\sigma\mu} + \Sigma_a(F_{Ca}{}^{1}{}_{\sigma\mu} + F_{Ca}{}^{2}{}_{\sigma\mu}) + iF_{B}{}^{1}{}_{\sigma\mu} + iF_{B}{}^{2}{}_{\sigma\mu} +\\ &\quad + \partial_\mu H^{1\beta}{}_{\sigma\beta} - \partial_\beta H^{1\beta}{}_{\sigma\mu} + H^{1\gamma}{}_{\beta\sigma}H^{1\beta}{}_{\gamma\mu} - H^{1\gamma}{}_{\mu\sigma}H^{1\beta}{}_{\gamma\beta} +\\ &\quad + \partial_\mu H^{2\beta}{}_{\sigma\beta} - \partial_\beta H^{2\beta}{}_{\sigma\mu} + H^{2\gamma}{}_{\beta\sigma}H^{2\beta}{}_{\gamma\mu} - H^{2\gamma}{}_{\mu\sigma}H^{2\beta}{}_{\gamma\beta} + H^{1\gamma}{}_{\beta\sigma}H^{2\beta}{}_{\gamma\mu} - H^{1\gamma}{}_{\mu\sigma}H^{2\beta}{}_{\gamma\beta} +\\ &\quad + H^{2\gamma}{}_{\beta\sigma}H^{1\beta}{}_{\gamma\mu} - H^{2\gamma}{}_{\mu\sigma}H^{1\beta}{}_{\gamma\beta} \end{aligned}$$

An additional Ricci-like tensor is

$$H_{\sigma_\mu} = H^{\beta}{}_{\sigma\beta\mu} \tag{12.17}$$

The curvature scalar is

$$\begin{aligned} R' = g^{\sigma\mu}R'_{\sigma\mu} &= + \partial^\sigma H^{1\beta}{}_{\sigma\beta} - \partial_\beta H^{1\beta}{}_{\sigma}{}^{\sigma} + H^{1\gamma}{}_{\beta\sigma}H^{1\beta}{}_{\gamma}{}^{\sigma} - H^{1\gamma}{}_{\mu\sigma}H^{1\beta}{}_{\gamma\beta} + \partial^\sigma H^{2\beta}{}_{\sigma\beta} - \partial_\beta H^{2\beta}{}_{\sigma}{}^{\sigma} +\\ &\quad + H^{2\gamma}{}_{\beta\sigma}H^{2\beta}{}_{\gamma}{}^{\sigma} - H^{2\gamma\sigma}{}_{\sigma}H^{2\beta}{}_{\gamma\beta} + H^{1\gamma}{}_{\beta\sigma}H^{2\beta}{}_{\gamma}{}^{\sigma} - H^{1\gamma\sigma}{}_{\sigma}H^{2\beta}{}_{\gamma\beta} + H^{2\gamma}{}_{\beta\sigma}H^{1\beta}{}_{\gamma}{}^{\sigma} -\\ &\quad - H^{2\gamma\sigma}{}_{\sigma}H^{1\beta}{}_{\gamma\beta} \end{aligned} \tag{12.18}$$

12.2 Vector Boson and Graviton Lagrangian Terms

Our higher derivative SU(4) vector boson and gravitational parts of the Lagrangian of the Unified SuperStandard Theory (with Generation, Layer, Higgs sector and Faddeev-Popov gauge term sectors not displayed here) are:

$$\mathcal{L} = \mathrm{Tr}\ \sqrt{g}[MD_\nu R''^{1}{}_{\sigma\mu}D^\nu R''^{2\sigma\mu} + aR'^{1}{}_{\sigma\mu}R'^{2\sigma\mu} + bR' + cg^{\sigma\mu}g^{2}{}_{\sigma\mu} + c'g^{2\sigma\mu}g^{2}{}_{\sigma\mu} - \\ - dA_{SU(3)}{}^{2}{}_{\mu}A_{SU(3)}{}^{2\mu}] \qquad (12.19)$$

where M, a, b, c, c', and d are constants, and $R''^{i}{}_{\sigma\mu}$ for i = 1, 2 determined above.[97]

This higher derivative Lagrangian is canonical and maintains the locality of the theory. It does entail a modest modification in the derivation of the Euler-Lagrange equations of motion. It also requires the use of principal value propagators rather than ordinary Feynman propagators for gluon and graviton interactions.

Thus the Strong Interaction sector, and the Gravitation sector are effectively Action-at-a-Distance theories that are similar in spirit to Wheeler-Feynman Electrodynamics. The two U(1) Electromagnetic sectors, the Generation group U(4) gauge field sector, the Layer group U(4) gauge field sector, the two SU2) Weak sectors, and the spinor connection sector may, or may not, be Action-at-a-Distance fields. They are not constrained to be Action-at-a-Distance by the present considerations.

Since we wish to apply our theory cosmologically, and within hadrons, where the gravitational spinor connections are negligible due to the smallness of the gravitational constant G and the 'smallness' of B spin on the cosmological scale, we set $F^{1}{}_{\nu\mu} = F^{2}{}_{\nu\mu} = 0$, and in view of the breaking of SU(4) to SU(3)⊗U(1), we find eq. 12.19 becomes[98]

$$\mathcal{L} = \mathrm{Tr}\ \sqrt{g}[MD_\nu(R'^{1}{}_{SU(3)\sigma\mu} + R^{1}{}_{\sigma\mu})D^\nu(R'_{SU(3)}{}^{2\sigma\mu} + R^{2\sigma\mu}) + \\ + aR'^{1}{}_{\sigma\mu}R'^{2\sigma\mu} + bR' + cg^{\sigma\mu}g^{2}{}_{\sigma\mu} + c'g^{2\sigma\mu}g^{2}{}_{\sigma\mu} - dA_{SU(3)}{}^{2}{}_{\mu}A_{SU(3)}{}^{2\mu}] \qquad (12.20)$$

---

[97] One may ask why $R''^{1}{}_{\sigma\mu}$ and $R''^{2}{}_{\sigma\mu}$ appear in the first term of the lagrangian, and not in other interaction terms. We believe the primary reason is: "The extended vierbein $l^{\mu ai}(x)$ can be viewed as located at a point in a higher dimensional complex-valued space.

$$l^{\mu ai}(x) = (\partial\xi_X{}^{ai}(x)/\partial x_\mu)_{X=h(x)}$$

where $\xi_X{}^{ai}$ is a set of locally inertial coordinates located at point X, and x = h(x) is a 4-dimensional point in a tangent subspace of the higher dimensional space:

$$X = h(x)$$

The relation between complex 4-dimensional coordinates x and the higher dimensional coordinates X is an embedding of a 4-dimensional surface within the higher dimensional complex space when account is taken of the range of possible x values. We have considered such embeddings in Blaha (2015a), and in earlier books, and developed a theory of a higher dimensional complex-valued space (the *Megaverse*) that contains our universe and probably many other universes." Thus SU(4) and Gravitation have a special role in our particle dynamics based on geometry. The second practical reason is the experimental reality that the Strong Interaction and Gravitation are known to have 'anomalous' features that will be seen to be remedied by these insertions while the other interactions are 'conventional.'

[98] The constants have the dimensions: M has the dimension of inverse mass squared, b has dimension mass squared, a is dimensionless, c and c' have dimension mass, and d has dimension mass squared.

Since there are no strong interaction fields in 'empty' space and gravity is negligible within hadrons,[99] we can drop the interaction terms between the Strong interaction and the Gravity interaction.[100] However, we cannot drop the interaction terms amongst Electromagnetism, the Weak interaction, the Strong Interaction, the Generation group U(4) interaction, and the U(4) Layer groups interactions. The interaction terms between Electromagnetism and Gravitation are important cosmologically.

The above Lagrangian terms can therefore be expressed as:[101]

$$\mathcal{L} = \mathcal{L}_E + \mathcal{L}_{SU(2)} + \mathcal{L}_{SU(3)} + \mathcal{L}_U + \mathcal{L}_V + \mathcal{L}_S + \mathcal{L}_\Theta + \mathcal{L}_G + \mathcal{L}_{int} \tag{12.21}$$

where taking traces of $\mathcal{L}$s terms is understood, and with coupling constants not displayed below to avoid clutter,

$$\mathcal{L}_E = \mathrm{Tr}\ \sqrt{g}\{M\{[\partial_v + i(A_E{}^1{}_v + A_E{}^2{}_v)]F^1{}_{E\sigma\mu}[\partial^v + i(A_E{}^{1v} + A_E{}^{2v})]F^2{}_E{}^{\sigma\mu}\} + aF_E{}^1{}_{\sigma\mu}F_E{}^{2\sigma\mu}\} \tag{12.22}$$

$$\mathcal{L}_{SU(2)} = \mathrm{Tr}\ \sqrt{g}[aF_W{}^1{}_{\sigma\mu}F_W{}^{2\sigma\mu}]$$

$$\mathcal{L}_{SU(3)} = \mathrm{Tr}\ \sqrt{g}\{M[\partial_v + i(A_{SU(3)}{}^1{}_v + A_{SU(3)}{}^2{}_v)]F_{SU(3)}{}^1{}_{\sigma\mu}[\partial^v + i(A_{SU(3)}{}^{1v} + A_{SU(3)}{}^{2v})]F_{SU(3)}{}^{2\sigma\mu} + aF_{SU(3)}{}^1{}_{\sigma\mu}F_{SU(3)}{}^{2\sigma\mu} - dA_{SU(3)}{}^2{}_\mu A_{SU(3)}{}^{2\mu}\}$$

$$\mathcal{L}_U = \mathrm{Tr}\ \sqrt{g}[aF_U{}^1{}_{\sigma\mu}F_U{}^{2\sigma\mu}]$$

$$\mathcal{L}_V = \mathrm{Tr}\ \sqrt{g}[aF_V{}^1{}_{\sigma\mu}F_V{}^{2\sigma\mu}]$$

$$\mathcal{L}_G = \mathrm{Tr}\ \sqrt{g}[MD_vR^1{}_{\sigma\mu}D^vR^{2\sigma\mu} + aR^1{}_{\sigma\mu}R^{2\sigma\mu} + bg^{\sigma\mu}(R^{1\beta}{}_{\sigma\beta\mu} + R^{2\beta}{}_{\sigma\beta\mu}) + cg^{\sigma\mu}g^2{}_{\sigma\mu} + c'g^{2\sigma\mu}g^2{}_{\sigma\mu}]$$
$$= \mathrm{Tr}\ \sqrt{g}[MD_vR^1{}_{\sigma\mu}D^vR^{2\sigma\mu} + aR^1{}_{\sigma\mu}R^{2\sigma\mu} + bH + cg^{\sigma\mu}g^2{}_{\sigma\mu} + c'g^{2\sigma\mu}g^2{}_{\sigma\mu}]$$

$$\mathcal{L}_{int} = \mathcal{L} - (\mathcal{L}_E + \mathcal{L}_{SU(2)} + \mathcal{L}_{SU(3)} + \mathcal{L}_U + \mathcal{L}_V + \mathcal{L}_G) \tag{12.23}$$

with appropriate sums over layers. Thus $\mathcal{L}_{SU(3)}$, $\mathcal{L}_{SU(2)}$, $\mathcal{L}_E$, $\mathcal{L}_U$,, $\mathcal{L}_V$, and parts of $\mathcal{L}_{int}$ are the dominant interactions within hadrons, and $\mathcal{L}_G$, $\mathcal{L}_E$ and parts of $\mathcal{L}_{int}$ are the dominant interactions in space within the framework of this discussion.

The $D_vR^1{}_{\sigma\mu}$ and $D^vR^{2\sigma\mu}$ terms have the form:

$$D_vR^i{}_{\sigma\mu} = +\,\partial_vR^i{}_{\sigma\mu} - H^{1\beta}{}_{\sigma v}R^i{}_{\beta\mu} - H^{1\beta}{}_{v\mu}R^i{}_{\sigma\beta} \tag{12.24}$$

for i = 1, 2.

The discussion continues in chapters 13 and 14.

---

[99] We show gravity weakens at very short distances using our Two-Tier Quantum Field Theory formalism. See Appendix A, and Blaha (2003) and (2005a) among other books by the author.

[100] We also drop the Dark sector part.

[101] We only consider the gauge field lagrangian terms.

13. The Strong Interaction Sector and the Higgs-Confinement Dichotomy

This chapter considers the Lagrangian formulation of SU(3) interactions in chapter 12 to quark confinement and the dichotomy between the Higgs Mechanism and Quark Confinement.[102]

13.1 Higgs Mechanism – Quark Confinement Dichotomy

Chapter 9 applied the two time coordinates formalism to separate symmetries into subluminal and superluminal categories. It placed SU(3)⊗U(1), which is SU(4) before splitting, in the subluminal category. It placed SU(2)⊗U(1), the ElectroWeak interactions, in the superluminal category. The SU(2)⊗U(1) placement as superluminal corresponds directly with the Higgs bosons tachyonic nature due to their negative masses squared in the ElectroWeak Langrangian. The question arises: Is it a coincidence? It appears not as we will see.

The deeper view of the Higgs Mechanism contrasted with the confining Strong interaction suggests that there is a Dichotomy in the sense that the Higgs Mechanism and the confining Strong Interaction are alternatives: one can have one or the other.

The Higgs Mechanism uses a tachyonic scalar particle implementation for symmetry breaking because it works and because it is superluminal. The Strong Interaction chooses a non-Higgs confinement because it is not tachyonic – it is subluminal. See chapter 9.

A simple example illustrating the dichotomy starts with the PseudoQuantum scalar field equations:

$$\Box\varphi_1 - m^2\varphi_1 + g\varphi_1^{\,4} = 0 \qquad (13.1)$$

$$\Box\varphi_2 - m^2\varphi_2 + g\varphi_2^{\,4} = 0 \qquad (13.2)$$

with tachyonic negative mass squared. Both the type 1 and type 2 fields have the same field equation as we have seen in earlier work.

On the other hand, the prototype Strong Interaction PseudoQuantum scalar field equations:

$$\Box\varphi_1 \pm m^2\varphi_2 = 0 \qquad (13.3)$$

$$\Box\varphi_2 = J \qquad (13.4)$$

with the quartic interaction dropped for simplicity yield

$$\Box^2\varphi_1 \pm m^2 J = 0 \qquad (13.5)$$

Choosing the minus sign to have a positive J source term: results in

[102] Sections 13.3 – 13.6 appeared previously in Blaha (2018e).

$$\Box^2\varphi_1 = m^2 J \qquad (13.6)$$

which specifies a confining linear r potential when redone as a gauge field theory.[103]

Note the minus sign places eqs. 13.1 and 13.3 on the same footing with the difference that eq. 13.1 is tachyonic but eq.13.3 is not tachyonic due to the appearance of the type 2 field in the eq. 13.3 mass term. "Tachyonicism" is avoided in order to obtain Strong interaction confinement.

Thus we see the *Higgs – Confinement Dichotomy*. The superluminal-subluminal two-times split of su(1, 1) and F-Theory (a SuperString[104] theory) in chapter 9 leads both to the split of symmetries in The Standard Model and UST, and to the Dichotomy between the Higgs Mechanism and the Confinement Strong Interaction.

These new successes, which explicate previously not understood phenomena, indicate that the superluminal-subluminal categorizations of chapter 9 are evidenced in physical phenomena. The structure of Standard Model symmetries: SU(2)⊗U(1)⊗SU(3) is now understandable. The relation and differentiation between the Higgs Mechanism and Strong Interaction Confinement is now explained.

13.2 Quark Confinement Potential

The confinement feature of the Strong Interaction has implications for the spectrum of bound quark states.

Its linear r potential, which is believed to lead to quark confinement, has an interesting unremarked feature: when used in relativistic Dirac equations to calculate the energies of multi-quark bound states. There is a V^2 term due to spinor arithmetic that can cause a linear r potential term to become similar to a harmonic oscillator r^2 potential term. The result is a spectrum with some harmonic oscillator-like features.[105]

13.3 Strong Interaction Lagrangian Terms

In the remainder of this chapter we describe some of the implications of the Strong Interaction sector of our theory described in chapter 12. We will find a linear quark potential and quark confinement follows.

The flat space-time Strong Interaction gauge field Lagrangian terms are

$$\begin{aligned}\mathcal{L}_{SU(3)} = \text{Tr}\,\sqrt{g}\{M[\partial_\nu + ig_1(A_{SU(3)}{}^1{}_\nu + A_{SU(3)}{}^2{}_\nu)]F_{SU(3)}{}^1{}_{\sigma\mu}[\partial^\nu + i\,g_1(A_{SU(3)}{}^{1\nu} + \\ + A_{SU(3)}{}^{2\nu})]F_{SU(3)}{}^{2\sigma\mu} + aF_{SU(3)}{}^1{}_{\sigma\mu}F_{SU(3)}{}^{2\sigma\mu} - dA_{SU(3)}{}^2{}_\mu A_{SU(3)}{}^{2\mu}\}\end{aligned} \qquad (13.7)$$

Dropping the SU(3) subscript for added clarity and adding color fermion terms we obtain

---

103 See Stephen Blaha, "New Framework for Gauge Field Theories", IL Nuovo Cimento **49A**, 113 (1979) for an SU(3) gauge theory version with confinement.

104 One qonders whether physical SuperString theories have a natural set of frequencies that are powers of two as we see in HyperCosmos theory.

105 In Blaha (2022d) the author points out the Dirac hydrogen atom equation has a $Z^2\alpha^2/r^2$ term[105] reflecting the V^2 type of term where $V = Z\alpha/r$.

$$\mathcal{L}_{SU(3)} = \text{Tr}\,\{MD_\nu F^1{}_{\sigma\mu}D^\nu F^{2\sigma\mu} + aF^1{}_{\sigma\mu}F^{2\sigma\mu} - dA^2{}_\mu A^{2\mu}\} + \bar{\psi}[i\not\nabla + g_1(\not A^1 + \not A^2) - m]\psi \quad (13.8)$$

where (for j = 1, 2)

$$D_\nu F^j{}_{\sigma\mu} = \partial_\nu F^j{}_{\sigma\mu} + ig_1[A^1{}_\nu, F^j{}_{\sigma\mu}] + ig_1[A^2{}_\nu, F^j{}_{\sigma\mu}] \quad (13.9)$$

The Lagrangian is equal to eq. 17 of S. Blaha, Phys. Rev. D**11**, 2921 (1974) except for additional terms $MD_\nu F^1{}_{\sigma\mu}D^\nu F^{2\sigma\mu}$ and $[A^2{}_\varkappa, A^2{}_\mu]$; and the following changes in parameters:

$$a = -\tfrac{1}{2} \qquad\qquad d = \tfrac{1}{2}\lambda^2$$

Since that paper essentially contains a complete description of our Strong Interaction theory (modulo the additional terms) we refer the reader to it and to its predecessor paper referenced therein. There are a few additional changes required to bring the 1974-5 papers into agreement with our current theory:

1. Eq. 30 of the above referenced paper must be modified to

$$F^2{}_{\varkappa\mu} = \partial A^2{}_\mu/\partial x^\varkappa - \partial A^2{}_\varkappa/\partial x^\mu + \mathbf{ig_1[A^2{}_\varkappa, A^2{}_\mu]} + ig_1[A^1{}_\varkappa, A^2{}_\mu] + ig_1[A^2{}_\varkappa, A^1{}_\mu] \quad (13.10)$$

with the addition of the 'bolded' term $if[A^2{}_\varkappa, A^2{}_\mu]$. There is also a trivial change of notation of coupling constant from 'g_1' to 'f'.

2. Eqs. 6 and 18 should have the interaction term expanded to

$$g_1\not A^1 \rightarrow g_1(\not A^1 + \not A^2) \quad (13.11)$$

and similarly in eq. 20. Eqs. 38 – 41. They directly show that the additional interaction term leads to a gluon propagator[106] $\langle A^1 + A^2, A^1 + A^2\rangle = 2\langle A^1, A^2\rangle + \langle A^1, A^1\rangle$, and introduces a 1/r term in the potential part of the gluon propagator.

As a result the effective gluon propagator in the theory, *if the* $Mf^2D_\nu F^1{}_{\sigma\mu}D^\nu F^{2\sigma\mu}$ *term is neglected*, combines eqs. 38 and 39 to give the *short-distance*[107] gluon propagator between quarks:

$$g_{\mu\nu}\delta_{ab}P[\lambda^2/k^4 - 1/k^2] \quad (13.12)$$

up to a constant factor.[108]

---

[106] Eqs. 40-41 in the above referenced paper.

[107] We anticipate that the $Mf^2D_\nu F^1{}_{\sigma\mu}D^\nu F^{2\sigma\mu}$ term will affect the short-distance behavior of the inter-quark interaction. The equivalent term in the gravitation sector influences the long-distance form of the gravity potential and leads to a MoND-like behavior. See chapter 23.

[108] This propagator is taken in Principal value to avoid potential unitarity problems. This topic is described in detail in earlier papers and books by the author.

These changes *explicitly* lead to a Strong Interaction potential of the form

$$V(r) = -2f^2/r + f^2\lambda^2 r \qquad (13.13)$$

Naturally one can expect perturbative corrections to eq. 13.13 in higher order in f. However, as will be seen later, the apparent relative smallness of f suggests eq. 13.13 is a good approximation to the **short-distance, *inter-quark interaction.***

13.3.1 Canonical Equal Time Commutation Relations

The Euler-Lagrange equations of motion, eqs. 27 – 31 in the author's 1974 paper are modified most significantly by the $Mf^2D_\nu F^1{}_{\sigma\mu}D^\nu F^{2\sigma\mu}$ Lagrangian term in eq. 13.7. In order to follow the canonical method to obtain the contributions to the equations of motion of this higher derivative term we will use integration by parts and discard surface terms, as is usually done in quantum field theory. We should start with eq. 13.8. However, with a view towards perturbation theory which appears reasonable in view of the smallness of the strong interaction coupling constant $f^2/4\pi = 0.024$ seen below, we will abstract a quadratic expression in the fields from eq. 13.8 and then proceed to develop gluon propagators and the strong interaction potential. The 'free' Strong Interaction Lagrangian that we use is

$$\mathcal{L}_{SU(3)F} = \mathrm{Tr}\{MD_{F\nu}F_F{}^{1a}{}_{\sigma\mu}D_F{}^\nu F_F{}^{2a\sigma\mu} + aF_F{}^{1a}{}_{\sigma\mu}F_F{}^{2a\sigma\mu} - dA^{2a}{}_\mu A^{2a\mu}\} + $$
$$+ \bar{\psi}[i\nabla\!\!\!/ + f(A\!\!\!/^1 + A\!\!\!/^2) - m]\psi \qquad (13.14)$$

where

$$F^{1a}{}_{\mu\varkappa} = \partial A^{1a}{}_\mu/\partial x^\varkappa - \partial A^{1a}{}_\varkappa/\partial x^\mu \qquad (13.15)$$
$$F^{2a}{}_{\mu\varkappa} = \partial A^{2a}{}_\mu/\partial x^\varkappa - \partial A^{2a}{}_\varkappa/\partial x^\mu$$

and

$$D_{F\nu} = \partial_\nu \qquad (13.16)$$

The conjugate momenta to $A^{1a}{}_\mu$ and $A^{2a}{}_\mu$ are respectively

$$\pi^{1a}{}_\mu = \partial\mathcal{L}_{SU(3)F}/(\partial A^{1a}{}_\mu/\partial t) = aF_F{}^{2a\mu t} \qquad (13.17)$$
$$\pi^{2a}{}_\mu = \partial\mathcal{L}_{SU(3)F}/(\partial A^{2a}{}_\mu/\partial t) = aF_F{}^{1a\mu t}$$

The non-zero, equal time commutation relations are

$$[\pi^{ia}{}_\mu(\mathbf{x}, t), A^{jb}{}_\nu(\mathbf{y}, t)] = i(1 - \delta^{ij})\delta^{ab}\delta^{G(\mu\nu)}(\mathbf{x} - \mathbf{y}) \qquad (13.18)$$

where i and j label the PseudoQuantum fields, and G(μν) indicates the gauge[109] G and the associated index expressions, with

[109] Not the gravitational coupling constant.

$$\delta^{G(\mu\nu)}(\mathbf{x}-\mathbf{y}) = \int d^4k \exp(-ik\cdot x)b^G{}_{\mu\nu}(k)/(2\pi)^4 \qquad (13.19)$$

where $b^G{}_{\mu\nu}(k)$ is a polynomial in k with a δ-function factor restricting the integration over k.

13.3.2 Dynamical Equations

After performing partial integrations on the $MD_{F\nu}F_F{}^1{}_{\sigma\mu}D_F{}^\nu F_F{}^{2\sigma\mu}$ term (and discarding surface terms at 'infinity') the Euler-Lagrange dynamical equations (in the Landau gauge) due to independent variations with respect to $A^{1a}{}_\mu$ is

$$2M\Box^2A^{2a}{}_\mu - 2a\Box A^{2a}{}_\mu = -g_1\bar{\psi}T^a\gamma_\mu\psi \qquad (13.20)$$

and, with respect to $A^{2a}{}_\mu$, is

$$2M\Box^2A^{1a}{}_\mu - 2a\Box A^{1a}{}_\mu - 2dA^{2a}{}_\mu = -g_1\bar{\psi}T^a\gamma_\mu\psi \qquad (13.21)$$

where T^a is an SU(3) generator. Subtracting the equations we find

$$2M\Box^2A^{1a}{}_\mu - 2a\Box A^{1a}{}_\mu - 2M\Box^2A^{2a}{}_\mu + 2a\Box A^{2a}{}_\mu - 2dA^{2a}{}_\mu = 0$$

or

$$A^{2a}{}_\mu = [2M\Box^2 - 2a\Box + 2d]^{-1}[2M\Box^2A^{1a}{}_\mu - 2a\Box A^{1a}{}_\mu] \qquad (13.22)$$

with the result

$$\{2M\Box^2 - 2a\Box - 2d[2M\Box^2 - 2a\Box + 2d]^{-1}[2M\Box^2 - 2a\Box]\}A^{1a}{}_\mu = -g_1\bar{\psi}T^a\gamma_\mu\psi$$

or

$$\{2M\Box^2 - 2a\Box - 2d[2M\Box^2 - 2a\Box + 2d]^{-1}[2M\Box^2 - 2a\Box]\}A^{1a}{}_\mu = -g_1\bar{\psi}T^a\gamma_\mu\psi$$
$$\{2M\Box^2 - 2a\Box - 2d + 4d^2[2M\Box^2 - 2a\Box + 2d]^{-1}\}A^{1a}{}_\mu = -g_1\bar{\psi}T^a\gamma_\mu\psi \qquad (13.23)$$

Eq. 13.16 leads to the Principal Value (Feynman) propagator:

$$D^{11}{}_{\mu\nu}(x-y) = P\, -i\langle 0|T(A^1{}_\mu(x), A^1{}_\nu(y)|0\rangle$$
$$= P\int d^4k \exp(-ik\cdot x)b_{\mu\nu}(k)D_1(k)/(2\pi)^4 \qquad (13.24)$$

where $b_{\mu\nu}(k)$ is a Landau gauge polynomial in k, and

$$D_1(k) = \{2Mk^4 - 2ak^2 - 2d + 4d^2[2Mk^4 - 2ak^2 + 2d]^{-1}\}^{-1}$$
$$= [2Mk^4 - 2ak^2 + 2d](2Mk^4 - 2ak^2)^{-2}$$

Thus

$$D^{11}{}_{\mu\nu}(x-y) = P\int d^4k \exp(-ik\cdot x)b_{\mu\nu}(k)[2Mk^4 - 2ak^2 + 2d]/[(2\pi)^4(2Mk^4 - 2ak^2)^2]$$
$$= P\int d^4k \exp(-ik\cdot x)b_{\mu\nu}(k)[2Mk^4 - 2ak^2 + 2d]/[(2\pi)^4k^4(2Mk^2 - 2a)^2] \qquad (13.25)$$

indicating a linear potential r term as well as terms of lower powers in r and Yukawa-like terms with a mass of $(a/M)^{½}$. We will describe the resulting effective Strong Interaction potential in more detail later.

Eq. 13.11 leads to the other propagator:

$$D^{12}{}_{\mu\nu}(x - y) = P -i\langle 0|T(A^1{}_\mu(x), A^2{}_\nu(y)|0\rangle$$
$$= P \int d^4k \exp(-ik\cdot x) b_{\mu\nu}(k) D_2(k)/(2\pi)^4 \tag{13.26}$$

where

$$D_2(k) = [2Mk^4 - 2ak^2]^{-1} \tag{13.27}$$

Thus

$$D^{12}{}_{\mu\nu}(x - y) = P \int d^4k \exp(-ik\cdot x) b_{\mu\nu}(k)/[(2\pi)^4 k^2(2Mk^2 - 2a)] \tag{13.28}$$

indicating a 1/r potential term plus a Yukawa term with a mass of $(a/M)^{½}$.

Due to the form of the interaction with quarks (eq. 13.14) the total effective gluon interaction between quarks is

$$D^{tot}{}_{\mu\nu}(x - y) = D^{11}{}_{\mu\nu}(x - y) + 2D^{12}{}_{\mu\nu}(x - y)$$
$$= P \int d^4k \exp(-ik\cdot x) b_{\mu\nu}(k)\{[3Mk^4 - 3ak^2 + 2d]/[(2\pi)^4 k^4(2Mk^2 - 2a)^2]\} \tag{13.29}$$

13.4 Strong Interaction Potential

Eq. 13.29 leads to the form of the total Strong Interaction potential. We note that the $\mu = \nu = 0$ part of the Feynman propagator for transverse gluons has the form:

$$D^{tot}{}_{00}(x - y) = \ldots - \int d^4k\, V_{SI}(\mathbf{k}) \exp(-ik\cdot(x - y))/(2\pi)^4 = \ldots + V_{SI}(\mathbf{x} - \mathbf{y})\delta(x_0 - y_0) \tag{13.30}$$

where

$$V_{SI}(\mathbf{x}) = -\int d^3k \exp(i\mathbf{k}\cdot\mathbf{x}) V_{SI}(\mathbf{k})/(2\pi)^3$$

with

$$V_{SI}(\mathbf{k}) = [3M\mathbf{k}^4 + 3a\mathbf{k}^2 + 2d]/[\mathbf{k}^4(2M\mathbf{k}^2 + 2a)^2] \tag{13.31}$$
$$= (2M)^{-2}\{2d(M/a)^2/\mathbf{k}^4 + [3a(M/a)^2 - 4d(M/a)^3]/\mathbf{k}^2$$
$$+ 2d(M/a)^2/(\mathbf{k}^2 + a/M)^2 + [-3a(M/a)^2 + 4d(M/a)^3]/(\mathbf{k}^2 + a/M)\}$$

Letting

$$m_{SI} = (a/M)^{½} \tag{13.32}$$

we find

$$V_{SI}(\mathbf{k}) = (2a)^{-2}\{2d/\mathbf{k}^4 + [3a - 4dm_{SI}^{-2}]/\mathbf{k}^2 + 2d/(\mathbf{k}^2 + m_{SI}^2)^2 + [4dm_{SI}^{-2} - 3a]/(\mathbf{k}^2 + m_{SI}^2)\} \tag{13.33}$$

The constant, a, is dimensionless and of order 1. The constant M has the dimension of inverse mass squared. We anticipate M will be extremely large resulting in a very small gluon mass m_{SI}.

There are massless gluon terms that generate color confinement reducing the impact of the massive gluon terms to a negligible effect outside hadronic regions.

We also see that the value of the inverse of graviton masses is of the order of the average galactic radius (the average galactic radius is large) and thus generate a Modified Newtonian potential (MoND) as seen in chapter 14.

Substituting eq. 13.33 we obtain a sum of massless and Yukawa-like potentials. The Yukawa potential is

$$V_Y(\mathbf{r}) = \int d^3k \exp(i\mathbf{k}\cdot\mathbf{r})/[(2\pi)^3(\mathbf{k}^2 + m^2)] = \exp(-mr)/[4\pi r] \quad (13.34)$$

Thus we obtain

$$V_{SI}(\mathbf{r}) = -(2a)^{-2}\{2d(dV_Y(\mathbf{r})/dm^2)|_{m=0} + [3a - 4dm_{SI}^{-2}]/(4\pi r) - 2d(dV_Y(\mathbf{r})/dm^2|_{m=m_{SI}}) + [4dm_{SI}^{-2} - 3a]V_Y(\mathbf{r})|_{m=m_{SI}}\} \quad (13.35)$$

with the form

$$V_{SI}(\mathbf{r}) = \alpha_1 r + \alpha_2/r + \alpha_3 e^{-m_{SI}r}/(4\pi m_{SI}) + \alpha_4 e^{-m_{SI}r}/(4\pi r) \quad (13.36)$$

where the constants α_i are:

$\alpha_1 = -(2a)^{-2}d/(8\pi)$ (up to an infrared divergent constant) (13.37)
$\alpha_2 = -(2a)^{-2}[3a - 4dm_{SI}^{-2}]/(4\pi)$
$\alpha_3 = -(2a)^{-2}d$
$\alpha_4 = -(2a)^{-2}[4dm_{SI}^{-2} - 3a]$

Thus we find the form of the potential of eq. 13.36 is 1/r and linear terms plus Yukawa-like terms with very small mass m_{SI} – perhaps near zero. As a result the form of the effective potential is

$$V(r) = -2g^2/r + g^2\lambda^2 r \quad (13.38)$$

13.5 The Origin of the Linear Potential

The linear potential appears to have originated in a suggestion of Feynman in the Spring of 1974. This author proposed[110] a non-Abelian canonical gauge quantum field theory, which yielded a linear potential. These papers, which had 4th order dynamic equations for the gauge fields, showed how to avoid the problems previously associated with higher derivative theories by using principal-value gauge field propagators that were similar in concept to the action-at-a-distance propagators used by Feynman and Wheeler in the late 1940's to formulate action-at-a-distance Quantum Electrodynamics.

Thus a non-Abelian quantum field theory of the strong interaction yielding a linear potential was created. In parallel with this development, Professor Kenneth Wilson (later a Nobelist) was developing lattice gauge theory. Because lattice lines

---

[110] S. Blaha, Phys. Rev. D**10**, 4268 (July, 1974) and Phys. Rev. D**11**, 2921 (December, 1974).

focus the field of gauge boson, lattice gauge theory exhibited a linear potential as well between quarks. Thus it offered an alternative to our gauge theory. However, the linearity of the lattice potential was "built-in" by the lattice theory formulation and thus was an artifact of the lattice formulation. This approach, and other proposed approaches, all share the problem that the linear potential that they produce cannot be proven to truly be a consequence. Rather the linear potential is the "likely result."

On the other hand our higher dimensional theory produces the linear potential if the standard canonical rules of quantum field theory are followed with the proviso that gauge field propagators are principal-value propagators.

This author had several discussions with Professor Wilson in late 1974 in the author's office and while walking to lunch at the Cornell Faculty Club. Professor Wilson proposed possible flaws in the author's theory on almost a daily basis. The author was able to show these suggested flaws were not flaws but physically acceptable. The final discussion with Wilson ended with Wilson stating words to the effect, "Your theory may be a correct phenomenological approximation to my theory of the strong interaction and quark confinement. But my theory is the correct one. Your theory is only a phenomenology." In the forty plus years since this concluding discussion no one has proved that the conventional strong interaction theory truly has a linear potential and quark confinement although some approximations suggest it does.

In the absence of a demonstration of a linear potential in the standard strong interaction model we suggest our theory is the only viable alternative. *An experimentally completely correct phenomenology is a theory.*

14. Gravitational Potential on Three Distance Scales vs. The MOND Alternative

Recently additional evidence has apparently been found for deviations from Newtonian Gravity. One proposed explanation is a modification of Newtonian Gravity called MOND. It seems to be in accord with some recent astronomical findings. MOND requires a modification of classical and quantum theory. This chapter suggests a more moderate explanation of data deviations based on higher derivative Gravitation. It finds that there are three regions of gravity: short, intermediate, and far distances corresponding to nearby regions, galactic regions and intergalactic regions as found experimentally.

This chapter[111] describes some of the possible results of the unified Lagrangian terms in eq. 12.22. Here we pull together results on the gravitational potential that were previously found in Blaha (2016g), (2016h) and (2017a). We note that a new experimental study of 33,000 galaxies indicate that the gravitational potential at inter-galactic distances deviates significantly from the Newtonian potential G/r. In Blaha (2017a) we showed that such a deviation occurs in our theory. This experimental result was not known at the time of its writing.[112]

In this chapter we will develop the gravitation sector.[113] We will see that the theory has higher derivative dynamic equations but unlike the Strong Interaction sector (next chapter) which yields color (quark-gluon) confinement, the gravitation dynamic equations do not have confinement – but do yield a modified form of gravity at various intermediate distances of the order of the average galactic radius. The modification of gravity implied by our theory is consistent with the need for Dark Matter described in our earlier Theory of Everything books (such as Blaha (2015a) and (2016c)).[114] It is also consistent with a MOND theory[115] with the addition of sixth order derivatives in the Gravitation sector Lagrangian in a manner consistent with the higher order terms appearing in the Strong Interaction sector of the unified theory. Thus our alternative to

---

[111] Most of this chapter appears in Blaha (2017a) and earlier books by the author.

[112] The gravitational potential was found to be greater than G/r at inter-galactic distances in a survey of 33,000 galaxies by M. Brouwer and colleagues at the Leiden Observatory (The Netherlands) in an announcement on December 18, 2016.

[113] There are other higher derivative theories – some with two metrics, and some with a metric plus vector plus scalar field formulation. The present work is based on a unification of Strong and Gravity sectors and a totally different formalism. Some significant references are: M. Milgrom, Phys. Rev. **D80**, 123536 (2009); C. Skordis et al, Phys. Rev. Lett. **96**, 011301 (2006); R. H. Sanders, Astrophysical Journal **480**, 492 (1997); and references therein; J. D. Bekenstein, Phys. Rev. **D70**, 083509 (2004) and references therein; J-P. Bruneton, Phys. Rev. **D76**, 124012 (2007) with a higher derivative gravity and metric, vector, scalar fields. See also references within these articles.

[114] I. Ferreras et al, Phys. Rev. Lett., **96**, 011301 (2006) shows the need for both Dark Matter and MOND based on studies of astrophysical data.

[115] Our gravity theory has aspects that are very similar to the MOND theories described in A. Balakin et al, Phys. Rev. **D70**, 064027 (2004); H-S Zhao et al, Phys. Rev. **D82**, 103001 (2010); and references therein. However our approach is very different.

MOND does not require a major change in Mechanics, quantum theory, or General Relativity (modulo higher order derivatives). The consistency between the need for higher order derivatives in both the Strong Interaction and Gravitation sectors is encouraging.

The developments in this chapter are a generalization with higher order derivative terms of the unified theory presented in chapter 6 of Blaha (2016d).

14.1 Gravitation Sector Dynamic Equations

Our gravity sector has two metric fields, $g_{\mu\nu}$ and $g^2{}_{\mu\nu}$ derived from the unified formalism described earlier. Some of the relevant gravitation equations found in sections 12.2 are:

$$H^\sigma{}_{\nu\mu} = \Gamma^\sigma{}_{\nu\mu} + \Gamma^{2\sigma}{}_{\nu\mu}$$

$$H^\beta{}_{\sigma\nu\mu} = \partial_\mu H^\beta{}_{\sigma\nu} - \partial_\nu H^\beta{}_{\sigma\mu} + H^\gamma{}_{\nu\sigma} H^\beta{}_{\gamma\mu} - H^\gamma{}_{\mu\sigma} H^\beta{}_{\gamma\nu}$$

$$H_{\sigma_\mu} = H^\beta{}_{\sigma\beta\mu}$$

$$H = g^{\sigma\mu} H_{\sigma\mu}$$
$$\mathcal{H} = R'^1 + R'^2$$

We use the gravitational sector Lagrangian:

$$\mathcal{L}_G = \sqrt{g}[M D_\nu R'^1{}_{G\sigma\mu} D^\nu R'^2{}_G{}^{\sigma\mu} + a R'^1{}_{G\sigma\mu} R'^2{}_G{}^{\sigma\mu} + b g^{\sigma\mu}(R'^1{}_G{}^\beta{}_{\sigma\beta\mu} + R'^2{}_G{}^\beta{}_{\sigma\beta\mu}) + c g^{\sigma\mu} g^2{}_{\sigma\mu} + \\ + e g^{2\sigma\mu} g^2{}_{\sigma\mu}]$$

$$\mathcal{L}_G = \sqrt{g}[M D_\nu R'^1{}_{G\sigma\mu} D^\nu R'^2{}_G{}^{\sigma\mu} + a R'^1{}_{G\sigma\mu} R'^2{}_G{}^{\sigma\mu} + bH + c g^{\sigma\mu} g^2{}_{\sigma\mu} + e g^{2\sigma\mu} g^2{}_{\sigma\mu}]$$

where

$$D_\nu V_\mu = (\partial_\nu + iF_\nu)V_\mu - H^\sigma{}_{\nu\mu} V_\sigma$$

$$= [g^\sigma{}_\mu \partial_\nu + i g^\sigma{}_\mu F_\nu - H^\sigma{}_{\nu\mu}] V_\sigma$$
$$= [g^\sigma{}_\mu \partial_\nu + i D^\sigma{}_{\mu\nu}] V_\sigma$$

where M, a, b, c, and d are constants:

$$a = 1/(2f) = 0.906$$

and

$$b = (8\pi G)^{-1}$$

where G is Newton's gravitational constant.

The Lagrangian dynamic equations that result are difficult. We consequently will examine the weak gravitation limiting case where we can approximate the two metrics with

$$g_{\mu\nu} \cong \eta_{\mu\nu} + h_{\mu\nu} \tag{14.1}$$
$$g^2_{\ \mu\nu} \cong \eta_{\mu\nu} + h^2_{\ \mu\nu} \tag{14.2}$$

where

$$|h_{\mu\nu}| << 1 \tag{14.3}$$
$$|h^2_{\ \mu\nu}| << 1$$

Using the relations

$$\partial_\mu h^\mu_{\ \nu} = \tfrac{1}{2}\partial_\nu h^\mu_{\ \mu} \tag{14.4}$$
$$\partial_\mu h^{2\mu}_{\ \ \nu} = \tfrac{1}{2}\partial_\nu h^{2\mu}_{\ \ \mu} \tag{14.5}$$

and neglecting higher order terms in $h_{\mu\nu}$ and $h^2_{\ \mu\nu}$ we find

$$\begin{aligned} R'^1_{\ \mu\nu} &= \tfrac{1}{2}[\Box h_{\mu\nu} - \partial_\mu\partial_\lambda h^\lambda_{\ \nu} - \partial_\nu\partial_\lambda h^\lambda_{\ \mu} + \partial_\nu\partial_\mu h^\lambda_{\ \lambda}] \\ &\cong \tfrac{1}{2}\Box h_{\mu\nu} \\ R'^1 &\cong \tfrac{1}{2}\Box h_\mu^{\ \mu} \end{aligned} \tag{14.6}$$

$$\begin{aligned} R'^2_{\ \mu\nu} &= \tfrac{1}{2}[\Box h^2_{\ \mu\nu} - \partial_\mu\partial_\lambda h^{2\lambda}_{\ \ \nu} - \partial_\nu\partial_\lambda h^{2\lambda}_{\ \ \mu} + \partial_\nu\partial_\mu h^{2\lambda}_{\ \ \lambda}] \\ &\cong \tfrac{1}{2}\Box h^2_{\ \mu\nu} \\ R'^2 &\cong \tfrac{1}{2}\Box h^{2\ \mu}_{\ \mu} \end{aligned} \tag{14.7}$$

with

$$D_\nu = \partial_\nu$$

upon neglecting higher order terms.

Substituting we find the *effective quadratic* part of the Lagrangian (in $h_{\mu\nu}$ and $h^2_{\ \mu\nu}$) is

$$\begin{aligned} \mathfrak{L}_G = \sqrt{g}[&M\partial_\nu R'^1_{\ G\sigma\mu}\partial^\nu R'^{2\ \sigma\mu}_{\ G} + a\Box h_{\sigma\mu}\Box h^{2\sigma\mu}/4 + \tfrac{1}{2}b(\partial_\alpha h^{\sigma\mu}\partial^\alpha h_{\sigma\mu} + \partial_\alpha h^{2\sigma\mu}\partial^\alpha h^2_{\ \sigma\mu}) + \\ &+ c(4 + \eta^{\sigma\mu}h^2_{\ \sigma\mu} + h^{\sigma\mu}\eta_{\sigma\mu} + h^{\sigma\mu}h^2_{\ \sigma\mu}) + \\ &+ e(2\eta^{\sigma\mu}h^2_{\ \sigma\mu} + h^{2\sigma\mu}h^2_{\ \sigma\mu}) + 1/4(h_{\mu\nu} + h^2_{\ \mu\nu})T^{\mu\nu}] \end{aligned} \tag{14.8}$$

Using partial integrations, we find the standard technique for determining the equations of motion from a Lagrangian for independent variations with respect to $h_{\mu\nu}$ and $h^2_{\ \mu\nu}$ yields

$$-M\Box^3 h^{2\mu\nu}/4 + a\Box^2 h^{2\mu\nu}/4 + \tfrac{1}{2}b\Box h^{\mu\nu} + c(\eta^{\mu\nu} + h^{2\mu\nu}) + 1/4T^{\mu\nu} = 0 \tag{14.9}$$

$$-M\Box^3 h^{\mu\nu}/4 + a\Box^2 h^{\mu\nu}/4 + \tfrac{1}{2}b\Box h^{2\mu\nu} + c(\eta^{\mu\nu} + h^{\mu\nu}) + 2e(\eta^{\mu\nu} + h^{2\mu\nu}) + 1/4T^{\mu\nu} = 0 \tag{14.10}$$

The term $c\eta^{\mu\nu}$ can be viewed as part of the total energy-momentum tensor $T'^{\mu\nu}$:

$$T'^{\mu\nu} = T^{\mu\nu} + 2c\eta^{\mu\nu} \tag{14.10a}$$

It plays a role similar to the Cosmological Constant. Subtracting the equations we find

$$M(\Box^3 h^{\mu\nu}/4 - \Box^3 h^{2\mu\nu}/4) + a\Box^2 h^{2\mu\nu}/4 - a\Box^2 h^{\mu\nu}/4 + ½b\Box h^{\mu\nu} - ½b\Box h^{2\mu\nu} + c(h^{2\mu\nu} - h^{\mu\nu}) - 2e(\eta^{\mu\nu} + h^{2\mu\nu}) = 0 \tag{14.11}$$

and thus

$$[-M\Box^3 + a\Box^2 - ½b\Box + (4c - 8e)]h^{2\mu\nu} = 8e\eta^{\mu\nu} - M\Box^3 h^{\mu\nu} + a\Box^2 h^{\mu\nu} - ½b\Box h^{\mu\nu} + 4ch^{\mu\nu} \tag{14.12}$$

Therefore we determine the metric equation for $h^{2\mu\nu}$

$$h^{2\mu\nu} = [-M\Box^3 + a\Box^2 - ½b\Box + (4c - 8e)]^{-1}[8e\eta^{\mu\nu} - M\Box^3 h^{\mu\nu} + a\Box^2 h^{\mu\nu} - ½b\Box h^{\mu\nu} + 4ch^{\mu\nu}] \tag{14.13}$$

Substituting in eq. 14.9 we obtain

$$[-M\Box^3/4 + a\Box^2/4 + c][-M\Box^3 + a\Box^2 - ½b\Box + (4c - 8e)]^{-1}[8e\eta^{\mu\nu} - M\Box^3 + a\Box^2 h^{\mu\nu} - ½b\Box h^{\mu\nu} + 4ch^{\mu\nu}] + \\ + ½b\Box h^{\mu\nu} + 1/4T'^{\mu\nu} = 0 \tag{14.14}$$

We now redefine the energy-momentum tensor with the result eq. L.14 becomes

$$\{[-M\Box^3/4 + a\Box^2/4 + c][-M\Box^3 + a\Box^2 - ½b\Box + (4c - 8e)]^{-1}[-M\Box^3 + a\Box^2 - ½b\Box + 4c] + \\ + ½b\Box\}h^{\mu\nu} = -1/4T'^{\mu\nu} \tag{14.15}$$

14.2 Gravity Potential

Assuming that we are dealing with non-relativistic matter we can calculate the gravity potential contribution from eq. 14.15

$$V_{G1}(\mathbf{x}) = -\int d^3k \exp(i\mathbf{k}\cdot\mathbf{x})V_{G1}(\mathbf{k})/(2\pi)^3 \tag{14.16}$$

where

$$V_{G1}(\mathbf{k}) = \{[M\mathbf{k}^6/4 + a\mathbf{k}^4/4 + c][M\mathbf{k}^6 + a\mathbf{k}^4 + ½b\mathbf{k}^2 + (4c - 8e)]^{-1}[M\mathbf{k}^6 + a\mathbf{k}^4 + ½b\mathbf{k}^2 + 4c] - \\ - ½b\mathbf{k}^2\}^{-1} \\ = [M\mathbf{k}^6 + a\mathbf{k}^4 + ½b\mathbf{k}^2 + (4c - 8e)]/\{[M\mathbf{k}^6/4 + a\mathbf{k}^4/4 - ½b\mathbf{k}^2 + c][M\mathbf{k}^6 + a\mathbf{k}^4 + ½b\mathbf{k}^2 + \\ + 4c] + 4eb\mathbf{k}^2\} \tag{14.17}$$

Similarly eq. 14.13 implies the other contribution to the total gravity potential is

$$V_{G2}(\mathbf{x}) = -\int d^3k \exp(i\mathbf{k}\cdot\mathbf{x})V_{G2}(\mathbf{k})/(2\pi)^3 \tag{14.18}$$

where

$$V_{G2}(\mathbf{k}) = \{[M\mathbf{k}^6 + a\mathbf{k}^4 + \tfrac{1}{2}b\mathbf{k}^2 + 4c + 8e]^{-1}[M\mathbf{k}^6 + a\,\mathbf{k}^4 + \tfrac{1}{2}b\mathbf{k}^2 + 4c - 8e]\}V_{G1}(\mathbf{k}) \quad (14.19)$$

The total gravity potential energy *for real-valued General Relativity* is thus

$$V^{tot}{}_{RG}(\mathbf{x}) = V_{G1}(\mathbf{x}) + V_{G2}(\mathbf{x}) \quad (14.20)$$

14.3 Solution for the Case of a Massless Graviton

Eqs. 14.16 - 14.17 can generate a massless graviton, which seems a requirement based on cosmological considerations, and also generate a pair of massive gravitons of very low mass. The massive gravitons generate a MOND-like potential that might explicate the anomalous gravitation effects seen at distances of the order of galactic dimensions.

If we set the 'Cosmological Constants' c = e = 0, then eq. 14.17 becomes

$$V_{G1}(\mathbf{k}) = 4/\{M\mathbf{k}^2[\mathbf{k}^4 + a\mathbf{k}^2/M - 2b/M]\} \quad (14.21)$$

The denominator of eq. 14.21 can be factored into the form

$$V_{G1}(\mathbf{k}) = 4/\{M\mathbf{k}^2[\mathbf{k}^2 + \tfrac{1}{2}m_{SI}^2 + \tfrac{1}{2}(m_{SI}^4 + 8bm_{SI}^2/a)^{1/2}][\mathbf{k}^2 + \tfrac{1}{2}m_{SI}^2 - \tfrac{1}{2}(m_{SI}^4 + 8bm_{SI}^2/a)^{1/2}]\} \quad (14.22)$$

where

$$m_A = (a/M)^{1/2}$$

Assuming $m_A^2 << 8b/a$, or $a^2/8 < 1/8 << bM = M(8\pi G)^{-1}$, which is reasonable since a is approximately one, we see

$$\begin{aligned} V_{G1}(\mathbf{k}) &\cong 4/\{M\mathbf{k}^2[\mathbf{k}^2 + (2bm_A^2/a)^{1/2}][\mathbf{k}^2 - (2bm_A^2/a)^{1/2}]\} \quad (14.23) \\ &= [2/(ba)]\{-2/\mathbf{k}^2 + 1/[\mathbf{k}^2 + (2bm_A^2/a)^{1/2}] + 1/[\mathbf{k}^2 - (2bm_A^2/a)^{1/2}]\} \end{aligned}$$

up to negligible terms.

From eq. 14.19 we see $V_{G2}(\mathbf{k}) = V_{G1}(\mathbf{k})$ if c = e = 0. Thus the total gravitational potential due to eq. 14.8 *for real-valued General Relativity* (in momentum space) is

$$\begin{aligned} V^{tot}{}_{RG}(\mathbf{k}) &= \tfrac{1}{2}(V_{G1}(\mathbf{k}) + V_{G2}(\mathbf{k})) \quad (14.24) \\ &= [2/(ba)]\{-2/\mathbf{k}^2 + 1/[\mathbf{k}^2 + (2bm_A^2/a)^{1/2}] + 1/[\mathbf{k}^2 - (2bm_A^2/a)^{1/2}]\} \\ &\cong 8\pi G\{-2/\mathbf{k}^2 + 1/[\mathbf{k}^2 + (2bm_A^2/a)^{1/2}] + 1/[\mathbf{k}^2 - (2bm_A^2/a)^{1/2}]\} \end{aligned}$$

and the coordinate space potential is[116]

$$V^{tot}{}_{RG}(\mathbf{r}) = -G/r + a_1Ge^{-m_Gr}/r + a_2G\cos(m_Gr)/r \quad (14.25)$$

where

$$m_G^2 = (2bm_A^2/a)^{1/2} = (2b/M)^{1/2} \quad (14.25a)$$

[116] Since the theory has higher order drivatives that could lead to unitarity problems Feynman propagators must be taken in Principal order. Since potentials are a part of Feynman propagators the potentials real value must be used.

$$a_1 = ½$$
$$a_2 = ½$$

where m_G is an extremely small mass.

Note that the real part of the expansion of the third term to third order yields

$$V^{tot}{}_{RG}(\mathbf{r}) \sim -G/r + a_1 G m_G{}^3 r^2 - a_2 G m_G{}^2 r + \text{constants} \quad (14.26)$$

with the resultant force

$$\mathbf{F} = \nabla V^{tot}{}_{RG}(\mathbf{r}) \sim G\mathbf{r}/r^3 + 2a_1 G m_G{}^3 \mathbf{r} - a_2 G m_G{}^2 \mathbf{r}/r + \ldots \quad (14.27)$$

This result is to be compared to the MOND force of A. Balakin et al, Phys. Rev. **D70**, 064027 (2004):

$$F = -\lambda Gm[M/r^2 - |\Pi_c|\, r/c^2] \quad (14.28)$$

and the vector form suggested by H-S Zhao et al, Phys. Rev. **D82**, 103001 (2010):

$$\partial\Phi/\partial\mathbf{r} = Gm\mathbf{r}/r^3 + (Gm)^{½}\mathbf{r}/r^2 \quad (14.29)$$

The resemblance to MOND estimates is clearly striking.

We chose the value of M such that the gravity potential at distances of the order of a 100,000 light years (the radius of the Andromeda galaxy) is increased by the Yukawa like terms due to the small value of m_G.[117] In addition the masses in the Yukawa-like parts of the Strong Interaction potential are extremely small but their effects are shielded so we find the effective Strong Interaction linear potential.

Thus our unified theory resolves the MOND gravity problem with a correct choice of M and maintains the Strong Interaction potential. A modification of Newton's law (F = ma) becomes unnecessary.

14.4 Intra-Solar System Distance Scale

Since m_G and m_A are presumably extremely small the gravitational potential at distances of up to at least several light years is approximately

$$V_{TOT}(\mathbf{r}) \cong -G/r \quad (14.30)$$

to well within feasible experimental limits.

[117] The third term in eq.14.25 is an oscillating Yukawa term that, because of the smallness of m_G, is slowly varying towards the end of a galaxy and thus could be well within observational error bounds. It appears that the real part of the third term is the contribution to the total gravitational potential using Principal value propagators.

14.5 Galactic Distance Scale

At distances of several tens of thousands of light years up to perhaps 100,000's of light years we find eq. 14.30 becomes approximately

$$V_{TOT}(\mathbf{r}) \cong -G/r + \tfrac{1}{2}G\cos(m_G r)/r \qquad (14.31)$$

Note that the approximate expansion of the terms in eq. 14.31 to third order yields

$$V_{TOT}(\mathbf{r}) \sim -G/r + a_1 G m_G{}^3 r^2 - a_1 G m_G{}^2 r + \text{constants} \qquad (14.32)$$

with the resultant force

$$\mathbf{F} = \nabla V_{TOT}(\mathbf{r})(\mathbf{r}) \sim G\mathbf{r}/r^3 + 2a_1 G m_G{}^3 \mathbf{r} - a_2 G m_G{}^2 \mathbf{r}/r + \ldots \qquad (14.33)$$

This result is to be compared to the MoND force of A. Balakin et al, Phys. Rev. **D70**, 064027 (2004):[118]

$$F = -\lambda Gm[M/r^2 \; - |\,\Pi_c\,|\; r/c^2]$$

and the vector form suggested by H-S Zhao et al, Phys. Rev. **D82**, 103001 (2010):

$$\partial\Phi/\partial\mathbf{r} \;=\; Gm\mathbf{r}/r^3 + (Gm)^{1/2}\mathbf{r}/r^2$$

Recent studies of 153 galaxies confirm the MoND discrepancy from Newtonian gravitation.[119]

14.6 Intergalactic Distance Scale

We can estimate the gravitational potential of $V_{TOT}(\mathbf{r})$ for large distances of the order of many hundreds of thousands of light years, and beyond, we see

$$V^{tot}{}_{RG}(\mathbf{r}) = -G/r + a_1 G e^{-m_G r}/r + a_2 G\cos(m_G r)/r \qquad (14.34)$$

provides a potential that is more positive than -G/r. This result is qualitatively consistent with some recent results in the study of open star clusters.[120]

Consequently we find a deeper potential than the galactic potential and thus a larger attractive gravitational force at inter-galactic distances in agreement with the 33,000 galaxy survey of M. Brouwer and colleagues.

14.7 Theoretical Agreement With Gravitational Data at all Known Distances

We conclude our unified theory agrees with known gravitational data at the three distance scales.

---

[118] The constant c above is the speed of light, and M is the mass (not the M used in our lagrangian equations.)

[119] S. S. McGaugh, F. Lelli, and J. M. Schombert, arXiv: 1609.0591 (2016).

[120] Jan Pflamm-Alternburg et al, Monthly Notices of the Royal Astronomical Society (October, 2022).

REFERENCES

Akhiezer, N. I., Frink, A. H. (tr), 1962, *The Calculus of Variations* (Blaisdell Publishing, New York, 1962).

Bjorken, J. D., Drell, S. D., 1964, *Relativistic Quantum Mechanics* (McGraw-Hill, New York, 1965).

Bjorken, J. D., Drell, S. D., 1965, *Relativistic Quantum Fields* (McGraw-Hill, New York, 1965).

Blaha, S., 1995, *C++ for Professional Programming* (International Thomson Publishing, Boston, 1995).

_______, 1998, *Cosmos and Consciousness* (Pingree-Hill Publishing, Auburn, NH, 1998 and 2002).

_______, 2002, *A Finite Unified Quantum Field Theory of the Elementary Particle Standard Model and Quantum Gravity Based on New Quantum Dimensions™ & a New Paradigm in the Calculus of Variations* (Pingree-Hill Publishing, Auburn, NH, 2002).

_______, 2004, *Quantum Big Bang Cosmology: Complex Space-time General Relativity, Quantum Coordinates,™ Dodecahedral Universe, Inflation, and New Spin 0, ½, 1 & 2 Tachyons & Imagyons* (Pingree-Hill Publishing, Auburn, NH, 2004).

______, *2005a, Quantum Theory of the Third Kind: A New Type of Divergence-free Quantum Field Theory Supporting a Unified Standard Model of Elementary Particles and Quantum Gravity based on a New Method in the Calculus of Variations* (Pingree-Hill Publishing, Auburn, NH, 2005).

______, 2005b, *The Metatheory of Physics Theories, and the Theory of Everything as a Quantum Computer Language* (Pingree-Hill Publishing, Auburn, NH, 2005).

______, 2005c, *The Equivalence of Elementary Particle Theories and Computer Languages: Quantum Computers, Turing Machines, Standard Model, Superstring Theory, and a Proof that Gödel's Theorem Implies Nature Must Be Quantum* (Pingree-Hill Publishing, Auburn, NH, 2005).

______, 2006a, *The Foundation of the Forces of Nature* (Pingree-Hill Publishing, Auburn, NH, 2006).

______, 2006b, *A Derivation of ElectroWeak Theory based on an Extension of Special Relativity; Black Hole Tachyons; & Tachyons of Any Spin.* (Pingree-Hill Publishing, Auburn, NH, 2006).

_______, 2007a, *Physics Beyond the Light Barrier: The Source of Parity Violation, Tachyons, and A Derivation of Standard Model Features* (Pingree-Hill Publishing, Auburn, NH, 2007).

______, 2007b, *The Origin of the Standard Model: The Genesis of Four Quark and Lepton Species, Parity Violation, the ElectroWeak Sector, Color SU(3), Three Visible Generations of Fermions, and One Generation of Dark Matter with Dark Energy* (Pingree-Hill Publishing, Auburn, NH, 2007).

______, *2008a, A Direct Derivation of the Form of the Standard Model From GL(16) (Pingree-Hill Publishing, Auburn, NH, 2008).*

______, 2008b, *A Complete Derivation of the Form of the Standard Model With a New Method to Generate Particle Masses Second Edition* (Pingree-Hill Publishing, Auburn, NH, 2008)

______, 2009, *The Algebra of Thought & Reality: The Mathematical Basis for Plato's Theory of Ideas, and Reality Extended to Include A Priori Observers and Space-Time Second Edition* (Pingree-Hill Publishing, Auburn, NH, 2009).

______, 2010a, *Operator Metaphysics: A New Metaphysics Based on a New Operator Logic and a New Quantum Operator Logic that Lead to a Mathematical Basis for Plato's Theory of Ideas and Reality* (Pingree-Hill Publishing, Auburn, NH, 2010).

______, 2010b, *The Standard Model's Form Derived from Operator Logic, Superluminal Transformations and GL(16)* (Pingree-Hill Publishing, Auburn, NH, 2010).

______, 2010c, *SuperCivilizations: Civilizations as Superorganisms* (McMann-Fisher Publishing, Auburn, NH, 2010).

______, 2011a, *21st Century Natural Philosophy Of Ultimate Physical Reality* (McMann-Fisher Publishing, Auburn, NH, 2011).

______, 2011b, *All the Universe! Faster Than Light Tachyon Quark Starships & Particle Accelerators with the LHC as a Prototype Starship Drive Scientific Edition* (Pingree-Hill Publishing, Auburn, NH, 2011).

______, 2011c, *From Asynchronous Logic to The Standard Model to Superflight to the Stars* (Blaha Research, Auburn, NH, 2011).

______, 2012a, *From Asynchronous Logic to The Standard Model to Superflight to the Stars volume 2: Superluminal CP and CPT, U(4) Complex General Relativity and The Standard Model, Complex Vierbein General Relativity, Kinetic Theory, Thermodynamics* (Blaha Research, Auburn, NH, 2012).

______, 2012b, *Standard Model Symmetries, And Four And Sixteen Dimension Complex Relativity; The Origin Of Higgs Mass Terms* (Blaha Reasearch, Auburn, NH, 2012).

______, 2013a, *Multi-Stage Space Guns, Micro-Pulse Nuclear Rockets, and Faster-Than-Light Quark-Gluon Ion Drive Starships* (Blaha Research, Auburn, NH, 2013).

______, 2013b, *The Bridge to Dark Matter; A New Sibling Universe; Dark Energy; Inflatons; Quantum Big Bang; Superluminal Physics; An Extended Standard Model Based on Geometry* (Blaha Reasearch, Auburn, NH, 2013).

______, 2014a, *Universes and Megaverses: From a New Standard Model to a Physical Megaverse; The Big Bang; Our Sibling Universe's Wormhole; Origin of the Cosmological Constant, Spatial Asymmetry of the Universe, and its Web of Galaxies; A Baryonic Field between Universes and Particles; Megaverse Extended Wheeler-DeWitt Equation* (Blaha Reasearch, Auburn, NH, 2014).

______, 2014b, *All the Megaverse! Starships Exploring the Endless Universes of the Cosmos Using the Baryonic Force* (Blaha Research, Auburn, NH, 2014).

______, 2014c, *All the Megaverse! II Between Megaverse Universes: Quantum Entanglement Explained by the Megaverse Coherent Baryonic Radiation Devices – PHASERs Neutron Star Megaverse Slingshot Dynamics Spiritual and UFO Events, and the Megaverse Microscopic Entry into the Megaverse* (Blaha Research, Auburn, NH, 2014).

______, 2015a, *PHYSICS IS LOGIC PAINTED ON THE VOID: Origin of Bare Masses and The Standard Model in Logic, U(4) Origin of the Generations, Normal and Dark Baryonic Forces, Dark Matter, Dark Energy, The Big Bang, Complex General Relativity, A Megaverse of Universe Particles* (Blaha Research, Auburn, NH, 2015).

______, 2015b, *PHYSICS IS LOGIC Part II: The Theory of Everything, The Megaverse Theory of Everything, U(4)⊗U(4) Grand Unified Theory (GUT), Inertial Mass = Gravitational Mass, Unified Extended Standard Model and a New Complex General Relativity with Higgs Particles, Generation Group Higgs Particles* (Blaha Research, Auburn, NH, 2015).

______, 2015c, *The Origin of Higgs ("God") Particles and the Higgs Mechanism: Physics is Logic III, Beyond Higgs – A Revamped Theory With a Local Arrow of Time, The Theory of Everything Enhanced, Why Inertial Frames are Special, Universes of the Mind* (Blaha Research, Auburn, NH, 2015).

______, 2015d, *The Origin of the Eight Coupling Constants of The Theory of Everything: U(8) Grand Unified Theory of Everything (GUTE), S^8 Coupling Constant Symmetry, Space-Time Dependent Coupling Constants, Big Bang Vacuum Coupling Constants, Physics is Logic IV* (Blaha Research, Auburn, NH, 2015).

______, 2016a, *New Types of Dark Matter, Big Bang Equipartition, and A New U(4) Symmetry in the Theory of Everything: Equipartition Principle for Fermions, Matter is 83.33% Dark, Penetrating the Veil of the Big Bang, Explicit QFT Quark Confinement and Charmonium, Physics is Logic V* (Blaha Research, Auburn, NH, 2016).

______, 2016b, *The Periodic Table of the 192 Quarks and Leptons in The Theory of Everything: The U(4) Layer Group, Physics is Logic VI* (Blaha Research, Auburn, NH, 2016).

______, 2016c, *New Boson Quantum Field Theory, Dark Matter Dynamics, Dark Matter Fermion Layer Mixing, Genesis of Higgs Particles, New Layer Higgs Masses, Higgs Coupling Constants, Non-Abelian Higgs Gauge Fields, Physics is Logic VII* (Blaha Research, Auburn, NH, 2016).

______, 2016d, *Unification of the Strong Interactions and Gravitation: Quark Confinement Linked to Modified Short-Distance Gravity; Physics is Logic VIII* (Blaha Research, Auburn, NH, 2016).

______, 2016e, *MoND: Unification of the Strong Interactions and Gravitation II, Quark Confinement Linked to Large-Scale Gravity, Physics is Logic IX* (Blaha Research, Auburn, NH, 2016).

______, 2016f, *CQ Mechanics: A Unification of Quantum & Classical Mechanics, Quantum/Semi-Classical Entanglement, Quantum/Classical Path Integrals, Quantum/Classical Chaos* (Blaha Research, Auburn, NH, 2016).

______, 2016g, *GEMS Unified Gravity, ElectroMagnetic and Strong Interactions: Manifest Quark Confinement, A Solution for the Proton Spin Puzzle, Modified Gravity on the Galactic Scale* (Pingree Hill Publishing, Auburn, NH, 2016).

______, 2016h, *Unification of the Seven Boson Interactions based on the Riemann-Christoffel Curvature Tensor* (Pingree Hill Publishing, Auburn, NH, 2016).

______, 2017a, *Unification of the Eleven Boson Interactions based on 'Rotations of Interactions'* (Pingree Hill Publishing, Auburn, NH, 2017).

______, 2017b, *The Origin of Fermions and Bosons, and Their Unification* (Pingree Hill Publishing, Auburn, NH, 2017).

______, 2017c, *Megaverse: The Universe of Universes* (Pingree Hill Publishing, Auburn, NH, 2017).

______, 2017d, *SuperSymmetry and the Unified SuperStandard Model* (Pingree Hill Publishing, Auburn, NH, 2017).

______, 2017e, *From Qubits to the Unified SuperStandard Model with Embedded SuperStrings: A Derivation* (Pingree Hill Publishing, Auburn, NH, 2017).

______, 2017f, *The Unified SuperStandard Model in Our Universe and the Megaverse: Quarks, ... ,* (Pingree Hill Publishing, Auburn, NH, 2017).

______, 2018a, *The Unified SuperStandard Model and the Megaverse SECOND EDITION A Deeper Theory based on a New Particle Functional Space that Explicates Quantum Entanglement Spookiness (Volume 1)* (Pingree Hill Publishing, Auburn, NH, 2018).

______, 2018b, *Cosmos Creation: The Unified SuperStandard Model, Volume 2, SECOND EDITION* (Pingree Hill Publishing, Auburn, NH, 2018).

______, 2018c, *God Theory (*Pingree Hill Publishing, Auburn, NH, 2018).

______, 2018d, *Immortal Eye: God Theory: Second Edition* (Pingree Hill Publishing, Auburn, NH, 2018).

______, 2018e, *Unification of God Theory and Unified SuperStandard Model THIRD EDITION* (Pingree Hill Publishing, Auburn, NH, 2018).

______, 2019a, *Calculation of: QED α = 1/137, and Other Coupling Constants of the Unified SuperStandard Theory* (Pingree Hill Publishing, Auburn, NH, 2019).

______, 2019b, *Coupling Constants of the Unified SuperStandard Theory SECOND EDITION* (Pingree Hill Publishing, Auburn, NH, 2019).

______, 2019c, *New Hybrid Quantum Big_Bang–Megaverse_Driven Universe with a Finite Big Bang and an Increasing Hubble Constant* (Pingree Hill Publishing, Auburn, NH, 2019).

______, 2019d, *The Universe, The Electron and The Vacuum* (Pingree Hill Publishing, Auburn, NH, 2019).

______, 2019e, *Quantum Big Bang – Quantum Vacuum Universes (Particles)* (Pingree Hill Publishing, Auburn, NH, 2019).

______, 2019f, *The Exact QED Calculation of the Fine Structure Constant Implies ALL 4D Universes have the Same Physics/Life Prospects* (Pingree Hill Publishing, Auburn, NH, 2019).

______, 2019g, *Unified SuperStandard Theory and the SuperUniverse Model: The Foundation of Science* (Pingree Hill Publishing, Auburn, NH, 2019).

______, 2020a, *Quaternion Unified SuperStandard Theory (The QUeST) and Megaverse Octonion SuperStandard Theory (MOST)* (Pingree Hill Publishing, Auburn, NH, 2020).

______, 2020b, *United Universes Quaternion Universe - Octonion Megaverse* (Pingree Hill Publishing, Auburn, NH, 2020).

______, 2020c, *Unified SuperStandard Theories for Quaternion Universes & The Octonion Megaverse* (Pingree Hill Publishing, Auburn, NH, 2020).

______, 2020d, *The Essence of Eternity: Quaternion & Octonion SuperStandard Theories* (Pingree Hill Publishing, Auburn, NH, 2020).

______, 2020e, *The Essence of Eternity II* (Pingree Hill Publishing, Auburn, NH, 2020).

______, 2020f, *A Very Conscious Universe* (Pingree Hill Publishing, Auburn, NH, 2020).

______, 2020g, *Hypercomplex Universe* (Pingree Hill Publishing, Auburn, NH, 2020).

______, 2020h, *Beneath the Quaternion Universe* (Pingree Hill Publishing, Auburn, NH, 2020).

______, 2020i, *Why is the Universe Real? From Quaternion & Octonion to Real Coordinates* (Pingree Hill Publishing, Auburn, NH, 2020).

______, 2020j, *The Origin of Universes: of Quaternion Unified SuperStandard Theory (QUeST); and of the Octonion Megaverse (UTMOST)* (Pingree Hill Publishing, Auburn, NH, 2020).

______, 2020k, *The Seven Spaces of Creation: Octonion Cosmology* (Pingree Hill Publishing, Auburn, NH, 2020).

______, 2020l, *From Octonion Cosmology to the Unified SuperStandard Theory of Particles* (Pingree Hill Publishing, Auburn, NH, 2020).

______, 2021a, *Pioneering the Cosmos* (Pingree Hill Publishing, Auburn, NH, 2021).

______, 2021b, *Pioneering the Cosmos II* (Pingree Hill Publishing, Auburn, NH, 2021).

______, 2021c, *Beyond Octonion Cosmology* (Pingree Hill Publishing, Auburn, NH, 2021).

______, 2021d, *Universes are Particles* (Pingree Hill Publishing, Auburn, NH, 2021).

______, 2021e, *Octonion-like dna-based life, Universe expansion is decay, Emerging New Physics* (Pingree Hill Publishing, Auburn, NH, 2021).

______, 2021f, *The Science of Creation New Quantum Field Theory of Spaces* (Pingree Hill Publishing, Auburn, NH, 2021).

______, 2021g, *Quantum Space Theory With Application to Octonion Cosmology & Possibly To Fermionic Condensed Matter* (Pingree Hill Publishing, Auburn, NH, 2021).

______, 2021h, *21st Century Natural Philosophy of Octonion Cosmology , and Predestination, Fate, and Free Will* (Pingree Hill Publishing, Auburn, NH, 2021).

______, 2021i, *Beyond Octonion Cosmology II : Origin of the Quantum; A New Generalized Field Theory (GiFT); A Proof of the Spectrum of Universes; Atoms in Higher Universes* (Pingree Hill Publishing, Auburn, NH, 2021).

______, 2021j, *Integration of General Relativity and Quantum Theory: Octonion Cosmology, GiFT, Creation/Annihilation Spaces CASe, Reduction of Spaces to a Few Fermions and Symmetries in Fundamental Frames* (Pingree Hill Publishing, Auburn, NH, 2021).

______, 2022a, *New View of Octonion Cosmology Based on the Unification of General Relativit and Quantum Theory* (Pingree Hill Publishing, Auburn, NH, 2022).

______, 2022b, *The Gold Dust Beneath Hypercomplex Cosmology* (Pingree Hill Publishing, Auburn, NH, 2022).

______, 2022c, *Passing Through Nature to Eternity: ProtoCosmos, HyperCosmos, Unified SuperStandard Theory* (Pingree Hill Publishing, Auburn, NH, 2022).

______, 2022d, *HyperCosmos Fractionation and Fundamental Reference Frame Based Unification: Particle Inner Space Basis of Parton and Dual Resonance Models* (Pingree Hill Publishing, Auburn, NH, 2022).

______, 2022e, *A New UniDimension ProtoCosmos and SuperString F-Theory Relation to the HyperCosmos* (Pingree Hill Publishing, Auburn, NH, 2022).

Eddington, A. S., 1952, *The Mathematical Theory of Relativity* (Cambridge University Press, Cambridge, U.K., 1952).

Fant, Karl M., 2005, *Logically Determined Design: Clockless System Design With NULL Convention Logic* (John Wiley and Sons, Hoboken, NJ, 2005).

Feinberg, G. and Shapiro, R., 1980, *Life Beyond Earth: The Intelligent Earthlings Guide to Life in the Universe* (William Morrow and Company, New York, 1980).

Gelfand, I. M., Fomin, S. V., Silverman, R. A. (tr), 2000, *Calculus of Variations* (Dover Publications, Mineola, NY, 2000).

Giaquinta, M., Modica, G., Souchek, J., 1998, *Cartesian Coordinates in the Calculus of Variations* Volumes I and II (Springer-Verlag, New York, 1998).

Giaquinta, M., Hildebrandt, S., 1996, *Calculus of Variations* Volumes I and II (Springer-Verlag, New York, 1996).

Gradshteyn, I. S. and Ryzhik, I. M., 1965, *Table of Integrals, Series, and Products* (Academic Press, New York, 1965).

Heitler, W., 1954, *The Quantum Theory of Radiation* (Claendon Press, Oxford, UK, 1954).

Huang, Kerson, 1992, *Quarks, Leptons & Gauge Fields 2nd Edition* (World Scientific Publishing Company, Singapore, 1992).

Jost, J., Li-Jost, X., 1998, *Calculus of Variations* (Cambridge University Press, New York, 1998).

Kaku, Michio, 1993, *Quantum Field Theory*, (Oxford University Press, New York, 1993).

Kirk, G. S. and Raven, J. E., 1962, *The Presocratic Philosophers* (Cambridge University Press, New York, 1962).

Landau, L. D. and Lifshitz, E. M., 1987, *Fluid Mechanics 2nd Edition*, (Pergamon Press, Elmsford, NY, 1987).

Misner, C. W., Thorne, K. S., and Wheeler, J. A., 1973, *Gravitation* (W. H. Freeman, New York, 1973).

Rescher, N., 1967, *The Philosophy of Leibniz* (Prentice-Hall, Englewood Cliffs, NJ, 1967).

Rieffel, Eleanor and Polak, Wolfgang, 2014, *Quantum Computing* (MIT Press, Cambridge, MA, 2014).

Riesz, Frigyes and Sz.-Nagy, Béla, 1990, *Functional Analysis* (Dover Publications, New York, 1990).

Sagan, H., 1993, *Introduction to the Calculus of Variations* (Dover Publications, Mineola, NY, 1993).

Sakurai, J. J., 1964, *Invariance Principles and Elementary Particles* (Princeton University Press, Princeton, NJ, 1964).

Weinberg, S., 1972, *Gravitation and Cosmology* (John Wiley and Sons, New York, 1972).

Weinberg, S., 1995, *The Quantum Theory of Fields Volume I* (Cambridge University Press, New York, 1995).

INDEX

12 dimension unification, 93
accumulation point, 9
Action-at-a-Distance, 113
amplitude trajectory α(s), 25
anti-spaces, 40
anti-universes, 40
baryonic force, 145
Big Bang, 42, 131, 132, 133, 143, 144
Bjorken, J. D., 131
Black Hole, 131
Blaha space number, 9
Boltzmann equation, 50
CASe, 33, 34, 37, 38, 39, 40
chaos, 50
charge, 14, 61, 62, 63
Charmonium, 133, 144
Classical, 133
color confinement, 51
complex, 43
Complex General Relativity, 132, 133
condensation, 43
Confinement, 11, 41, 51, 101, 105, 115, 116, 133, 144
Connection groups, 65, 67, 68, 70, 72
conservation law, 45
Cosmological Constant, 132
Cosmos, iv, 134, 144
coupling constants, 30, 52, 110
CQ Mechanics, 49
Creation, 134
Dark Energy, 42, 131
Dark Matter, 131, 132, 133
decay to lower resonances, 26
deep inelastic, 144
Deep Inelastic lepton-nucleon scattering, 7
Dichotomy, 11, 101, 115, 116
Dimensional Interaction, 43
dimensions, 43
divergences, 143
Dual Resonance, iv, 3, 7, 11, 136
ElectroWeak, 131
Entropy, 18, 23
equal time commutation relations, 46
Feynman propagators, 43, 53, 113
fine structure constant, 144
Fokker-Planck equation, 50
FRF, vii, xii, 1, 2, 58, 59, 80, 81, 82, 83, 84, 86, 89, 90, 97, 101, 102
F-Theory, iv, vii, xv, 1, 2, 93, 99, 100, 116, 136
Fundamental Reference Frame, iv, vii, 1, 11, 16, 19, 58, 75, 76, 78, 79, 80, 81, 82, 83, 84, 85, 89, 101, 102, 105, 136
Gambols, 7
gauge invariance, 44, 45
Gaussian, 43
General Relativistic transformations, 37, 39, 107
Generalized Field Theory, 29, 136
Generation group, 5, 57, 61, 113, 114. 133
ghost, 43
GiFT, iv, 29, 30, 34, 39, 136
gluon propagator, 117
Grand Unified Theory of Everything, 133
Gravitational Potential, 123
Gulliver's Travels, 1
Gupta-Bleuler gauge, 46
GUT, 133
harmonic oscillator, 143
Heisenberg Uncertainty Relation, 47
Heitler, 47

Higgs Mechanism, vii, xv, 11, 30, 52, 59, 77, 87, 101, 103, 105, 115, 116, 133, 143
Higgs particles, 30, 52, 143
higher derivative SU(3), xv
higher space, 94, 95, 97
Hydrogen, 9, 39
Hypercomplex Cosmology, 2, 85, 95, 107
HyperCosmos, iv, 10, 29, 39, 40, 41, 42, 43, 49, 51, 65, 70, 75, 76, 78, 79, 81, 83, 84, 85, 86, 89, 100, 103, 107, 136, 146
imaginary coordinates, 43
interactions, 143
Interdimensional Interaction, 43
inverted spectrum, 9
ISIS, 145
Jacobian, 44
Killing vector, 30, 34, 39, 42, 51
Klein-Gordon field, 52
lagrangian, 44, 45
lattice gauge theory, 7, 121
Layer group, 5, 57, 113, 114
Leibniz, 137
LHC, iv, 132
Lilliputians, 1
Limos, v, 1, 7, 8, 17
Linear Potential, 121
mass scale, 43
massless graviton, 127
Megaverse, iv, 1, 7, 50, 70, 93, 98, 108, 113, 134, 135, 146
Milgrom, 123
Modified Newtonian Potential, 127, 128, 129
MOND, vii, xv, 30, 123, 127, 128, 144
NEWMaxiverse, 72
NEWQUeST, 37, 71
NEWUTMOST, 65, 70, 71, 72
Octonion, 135
Octonion Cosmology, 135
paradoxes, 143
Parity Violation, 131
Parton Model, 11
partons, 11
perturbation theory, 41, 42, 43, 48, 53, 144
Plato's Theory of Ideas, 132
ProtoCosmos, iv, 39, 136, 146
PseudoQuantum Theory, vi, 10, 29, 30, 41
Quantum, 134, 143, 145
quantum computers, 143
Quantum Dimensions, 43, 131
Quantum Entanglement, 132
quantum fluctuations, 43
Quantum Gravity, 131
quark, 143
quark confinement, 42, 50
QUeST, v, vi, vii, ix, x, xi, xii, xiii, 3, 5, 6, 10, 11, 39, 55, 56, 57, 58, 59, 60, 65, 67, 68, 75, 77, 78, 89, 90, 91, 93, 99, 100, 102, 105, 135, 146
radiation gauge, 46
Regge pole trajectory, 9
Regge Trajectory, v, xv, 25
Relic Structure, 11
replication, 1, 76, 77, 80, 87
Ricci tensor, 112
Riemann-Christoffel curvature tensor, xv, 107
Robertson-Walker metric, 143
scalar field, 52
scaling, 9, 11, 144
sedenion, 34
$SO^{+}(1, 3)$, 75, 77, 78, 85, 100
Special Relativity, 131
species, 61, 62, 63
spin, 143
spinor connection, 113
Standard Model, v, vi, vii, ix, xii, xiii, xv, 3, 7, 10, 11, 30, 52, 59, 75, 76, 85, 86, 88, 89, 91, 93, 100, 103, 105, 108, 116, 131, 132, 133, 143, 144, 145
static reference frame, 84

String Theories, 11
Strong Interaction potential, 118, 120
Strong Interaction sector, 116
SU(3), 131
SU(4), 67
superluminal, 3, 85, 86, 87, 88, 89, 91
SuperStandard Model, iv, 134, 135
Superstring, 131
symmetry splitting, 101
Thermodynamics, 132
time coordinates, 93, 95, 99
Tower ProtoCosmos, v, x, xv, 8, 19, 21, 22, 23, 29
translational invariance, 45
Twistor Theory, 33
two time coordinates, ix, xv, 2, 3, 58, 85, 86, 93, 115
Two-Tier, 41, 42, 43, 44, 47, 48, 49, 52, 53
Two-Tier coordinates, 42
two-tier formalism, 44
Two-Tier Quantum Field Theory, 42
U(4), 132, 133
U(8), 133
UniDimension Model, 13
UniDimension ProtoCosmos, iv, v, ix, xv, 1, 7, 11, 13, 16, 17, 18, 23, 25, 27, 29, 136
unification, 33
Unification, 33, 133, 134, 136
unification space, 95
Unified SuperStandard Model, iv, 134, 135
unitarity, 44, 48, 49
universe wave function, 22, 27
UST, 5, 57, 146
Veneziano Amplitude, 7
Web of Galaxies, 132
Wheeler-DeWitt equation, 145
Wheeler-Feynman Electrodynamics, 113
Wilson, K., 121
Yukawa-like, 120, 121, 128

About the Author

Stephen Blaha is a well-known Physicist and Man of Letters with interests in Science, Society and civilization, the Arts, and Technology. He had an Alfred P. Sloan Foundation scholarship in college. He received his Ph.D. in Physics from Rockefeller University. He has served on the faculties of several major universities. He was also a Member of the Technical Staff at Bell Laboratories, a manager at the Boston Globe Newspaper, a Director at Wang Laboratories, and President of Blaha Software Inc. and of Janus Associates Inc. (NH).

Among other achievements he was a co-discoverer of the "r potential" for heavy quark binding developing the first (and still the only demonstrable) non-Aeolian gauge theory with an "r" potential; first suggested the existence of topological structures in superfluid He-3; first proposed Yang-Mills theories would appear in condensed matter phenomena with non-scalar order parameters; first developed a grammar-based formalism for quantum computers and applied it to elementary particle theories; first developed a new form of quantum field theory without divergences (thus solving a major 60 year old problem that enabled a unified theory of the Standard Model and Quantum Gravity without divergences to be developed); first developed a formulation of complex General Relativity based on analytic continuation from real space-time; first developed a generalized non-homogeneous Robertson-Walker metric that enabled a quantum theory of the Big Bang to be developed without singularities at t = 0; first generalized Cauchy's theorem and Gauss' theorem to complex, curved multi-dimensional spaces; received Honorable Mention in the Gravity Research Foundation Essay Competition in 1978; first developed a physically acceptable theory of faster-than-light particles; first derived a composition of extremums method in the Calculus of Variations; first quantitatively suggested that inflationary periods in the history of the universe were not needed; first proved Gödel's Theorem implies Nature must be quantum; provided a new alternative to the Higgs Mechanism, and Higgs particles, to generate masses; first showed how to resolve logical paradoxes including Gödel's Undecidability Theorem by developing Operator Logic and Quantum Operator Logic; first developed a quantitative harmonic oscillator-like model of the life cycle, and interactions, of civilizations; first showed how equations describing superorganisms also apply to civilizations. A recent book shows his theory applies successfully to the past 14 years of history and to *new* archaeological data on Andean and Mayan civilizations as well as Early Anatolian and Egyptian civilizations.

He first developed an axiomatic derivation of the form of The Standard Model from geometry – space-time properties – The Unified SuperStandard Model. It unifies all the known forces of Nature. It also has a Dark Matter sector that includes a Dark ElectroWeak sector with Dark doublets and Dark gauge interactions. It uses quantum coordinates to remove infinities that crop up in most

interacting quantum field theories and additionally to remove the infinities that appear in the Big Bang and generate inflationary growth of the universe. It shows gravity has a MOND-like form without sacrificing Newton's Laws. It relates the interactions of the MOND-like sector of gravity with the r-potential of Quark Confinement. The axioms of the theory lead to the question of their origin. We suggest in the preceding edition of this book it can be attributed to an entity with God-like properties. We explore these properties in "God Theory" and show they predict that the Cosmos exists forever although individual universes (or incarnations of our universe) "come and go." Several other important results emerge from God Theory such a functionally triune God. The Unified SuperStandard Theory has many other important parts described in the Current Edition of *The Unified SuperStandard Theory* and expanded in subsequent volumes.

Blaha has had a major impact on a succession of elementary particle theories: his Ph.D. thesis (1970), and papers, showed that quantum field theory calculations to all orders in ladder approximations could not give scaling deep inelastic electron-nucleon scattering. He later showed the eigenvalue equation for the fine structure constant α in Johnson-Baker-Willey QED had a zero at α = 1 not 1/137 by solving the Schwinger-Dyson equations to all orders in an approximation that agreed with exact results to 4th order in α thus ending interest in this theory. In 1979 at Prof. Ken Johnson's (MIT) suggestion he calculated the proton-neutron mass difference in the MIT bag model and found the result had the wrong sign reducing interest in the bag model. These results all appear in Physical Review papers. In the 2000's he repeatedly pointed out the shortcomings of SuperString theory and showed that The Standard Model's form could be derived from space-time geometry by an extension of Lorentz transformations to faster than light transformations. This deeper space-time basis greatly increases the possibility that it is part of THE fundamental theory. Recently, Blaha showed that the Weak interactions differed significantly from the Strong, electromagnetic and gravitation interactions in important respects while these interactions had similar features, and suggested that ElectroWeak theory, which is essentially a glued union of the Weak interactions and Electromagnetism, possibly modulo unknown Higgs particle features, be replaced by a unified theory of the other interactions combined with a stand-alone Weak interaction theory. Blaha also showed that, if Charmonium calculations are taken seriously, the Strong interaction coupling constant is only a factor of five larger than the electromagnetic coupling constant, and thus Strong interaction perturbation theory would make sense and yield physically meaningful results.

In graduate school (1965-71) he wrote substantial papers in elementary particles and group theory: The Inelastic E- P Structure Functions in a Gluon Model. Phys. Lett. B40:501-502,1972; Deep-Inelastic E-P Structure Functions In A Ladder Model With Spin 1/2 Nucleons, Phys.Rev. D3:510-523,1971; Continuum Contributions To The Pion Radius, Phys. Rev. 178:2167-2169,1969; Character Analysis of U(N) and SU(N), J. Math. Phys. 10, 2156 (1969); and The Calculation of the Irreducible Characters of the Symmetric Group in Terms of the

Compound Characters, (Published as Blaha's Lemma in D. E. Knuth's book: *The Art of Computer Programming Vols. 1 – 4*).

In the early 1980's Blaha was also a pioneer in the development of UNIX for financial, scientific and Internet applications: benchmarked UNIX versions showing that block size was critical for UNIX performance, developing financial modeling software, starting database benchmarking comparison studies, developing Internet-like UNIX networking (1982) and developing a hybrid shell programming technique (1982) that was a precursor to the PERL programming language. He was also the manager of the AT&T ten-year future products development database. His work helped lead to commercial UNIX on computers such as Sun Micros, IBM AIX minis, and Apple computers.

In the 1980's he pioneered the development of PC Desktop Publishing on laser printers and was nominated for three "Awards for Technical Excellence" in 1987 by PC Magazine for PC software products that he designed and developed.

Recently he has developed a theory of Megaverses – actual universes of which our universe is one – with quantum particle-like properties based on the Wheeler-DeWitt equation of Quantum Gravity. He has developed a theory of a baryonic force, which had been conjectured many years ago, and estimated the strength of the force based on discrepancies in measurements of the gravitational constant G. This force, operative in D-dimensional space, can be used to escape from our universe in "uniships" which are the equivalent of the faster-than-light starships proposed in the author's earlier books. Thus travel to other universes, as well as to other stars is possible.

Blaha also considered the complexified Wheeler-DeWitt equation and showed that its limitation to real-valued coordinates and metrics generated a Cosmological Constant in the Einstein equations.

The author has also recently written a series of books on the serious problems of the United States and their solution as well as a book on the decline of Mankind that will follow from current social and genetic trends in Mankind.

In the past twenty years Dr. Blaha has written over 80 books on a wide range of topics. Some recent major works are: *From Asynchronous Logic to The Standard Model to Superflight to the Stars*, *All the Universe!*, *SuperCivilizations: Civilizations as Superorganisms*, *America's Future: an Islamic Surge, ISIS, al Qaeda, World Epidemics, Ukraine, Russia-China Pact, US Leadership Crisis*, *The Rises and Falls of Man – Destiny – 3000 AD: New Support for a Superorganism MACRO-THEORY of CIVILIZATIONS From CURRENT WORLD TRENDS and NEW Peruvian, Pre-Mayan, Mayan, Anatolian, and Early Egyptian Data, with a Projection to 3000 AD*, and *Mankind in Decline: Genetic Disasters, Human-Animal Hybrids, Overpopulation, Pollution, Global Warming, Food and Water Shortages, Desertification, Poverty, Rising Violence, Genocide, Epidemics, Wars, Leadership Failure.*

He has taught approximately 4,000 students in undergraduate, graduate, and postgraduate corporate education courses primarily in major universities, and large companies and government agencies.

He developed a quantum theory, The Unified SuperStandard Theory (UST), which describes elementary particles in detail without the difficulties of conventional quantum field theory. He found that the internal symmetries of this theory could be exactly derived from an octonion theory called QUeST. He further found that another octonion theory (UTMOST) describes the Megaverse. It can hold QUeST universes such as our own universe. It has an internal symmetry structure which is a superset of the QUeST internal symmetries.

Recently he developed Octonion Cosmology. He replaced it with HyperCosmos theory, which has significantly better features. He developed a fractionalization process for dimensions, particles and symmetry groups. He also described transformation that reduced particles and dimensions to a far more compact form. He also developed a precursor theory ProtoCosmos that leads to the HyperCosmos.

www.ingramcontent.com/pod-product-compliance
Lightning Source LLC
LaVergne TN
LVHW071338210926
848486LV00003B/173